D1585801

An Introduction to Computer Aided Production Management

Stephen J. Childe

University of Plymouth
Plymouth, UK

CHAPMAN & HALL

London · Weinheim · New York · Tokyo · Melbourne · Madras

**Published by Chapman & Hall, 2–6 Boundary Row,
London SE1 8HN, UK**

Chapman & Hall, 2–6 Boundary Row, London SE1 8HN, UK

Chapman & Hall GmbH, Pappelallee 3, 69469 Weinheim, Germany

Chapman & Hall USA, 115 Fifth Avenue, New York, NY 10003, USA

Chapman & Hall Japan, ITP-Japan, Kyowa Building, 3F, 2-2-1
Hirakawacho, Chiyoda-ku, Tokyo 102, Japan

Chapman & Hall Australia, 102 Dodds Street, South Melbourne, Victoria
3205, Australia

Chapman & Hall India, R. Seshadri, 32 Second Main Road, CIT East,
Madras 600 035, India

First edition 1997

© 1997 Stephen J. Childe

Typeset in 10/12 pt Palatino by Saxon Graphics Ltd, Derby
Printed in Great Britain by St. Edmundsbury Press, Bury St. Edmunds,
Suffolk

ISBN 0 412 62010 3

A catalogue record for this book is available from the British Library

Library of Congress Catalog Card Number: 96-85395

∞ Printed on permanent acid-free text paper, manufactured in
accordance with ANSI/NISO Z39.48-1992 and ANSI/NISO Z39.48-1984
(Permanence of Paper).

Contents

Preface

WHO MIGHT READ THIS BOOK?

This book is written for students of manufacturing systems as well as people in industry who may need an explanation of some of the concepts of CAPM or who may be looking for a new idea. A high level of prior knowledge of manufacturing is not required, and the book attempts to explain production management approaches in simple language.

The book aims to provide the reader with a sufficient appreciation of computer aided production management to understand the basic operation of the systems which may be encountered in industry, to question their application and evaluate alternatives, to be able to discuss CAPM with experts in the field such as system vendors, and to contribute to the design and implementation of new systems.

The book is intended to proceed in a logical fashion from the start to the end, so the later chapters do assume some knowledge of the earlier ones. However, it is realized that the reader may need to dip into specific areas, so references are provided to ideas presented earlier in the text where appropriate, and to useful texts and papers.

WHY IS CAPM IMPORTANT?

Manufacturing is about performing processes on materials to make products which can be sold. If this can be done well enough to make a profit, it is of benefit to the owners, employees and customers of the manufacturing company, and ultimately adds to the wealth of the country's economy.

However, all manufacturing companies operate in a competitive environment, so their customers may be faced with several competing products. In order to make a sale, the manufacturing company must deliver the product as closely as possible to the customer's requirements while keeping production costs low enough to make a profit at a selling price that the customer will accept. This book aims to provide an explanation of the principal ways in which production is managed to achieve competitiveness.

The dimensions of competitiveness vary enormously between different products, different types of customer and in different regions. Competitiveness is made up of a complex mix of factors, including product design and performance, advertising and promotion, company reputation, customer service and after-sales service, besides price and availability. But no matter what the product, industry sector or customer, the competent management of production is essential for the competitiveness of the manufacturing business.

The control of manufacturing operations is one of the most important aspects of manufacturing industry. The correct regulation of manufacturing activities makes the difference between meeting and missing customer requirements. Effective stock control ensures that the items needed for manufacture are ready at the right time and without the company investing money in excess stocks which are available when they are not required, meaning that cash is not available to buy what is needed. Effective production control increases product quality by being able to identify, trace and correct any variations in performance, equipment or materials, without allowing any effect to detract from the final product. Production control also seeks to reduce manufacturing lead times so that less work is in progress at any instant, and new products can be brought to the market-place as quickly as possible. Production management is at the heart of manufacturing competitiveness.

Different approaches to production control have drawn attention to various areas, such as stock control, shop floor scheduling and material purchasing. These have led to a range of rules of thumb, common practices and computer systems. It is the purpose of this book to explain the main approaches and their application. Attention is also paid to the business context and to the implementation of new working methods.

The term 'Computer Aided Production Management' is widely used. It should be understood that not all aspects of production management can be computerized, nor should they be. This book is written from the point of view that computerized solutions should be available as an aid to production management – they can make possible operations which an army of human production controllers could never manage – but they also bring problems of their own, especially when a complex computer system is used in an attempt to control a complex factory.

Nowadays, most companies use some kind of computer support in their production management, but it is important to remember that good control comes from developing a good understanding of the way the factory works and making it as simple as possible to operate.

If the control problem can be simplified, even if only by splitting it into a set of smaller problems, the likely success of any new production management system, manual or computerized, is greatly increased. Thus the production system as a whole must be considered and, if necessary, the factory, its equipment and the organization must be redesigned for effective control. The production management system is seen as part of a whole 'manufacturing system', consisting of equipment, information, procedures and organizational structures. It is important to understand the whole system before attempting to deal with any part.

STRUCTURE

Chapter 1 looks at the competitive strategies that manufacturing businesses need to fulfil in order to satisfy the requirements of their owners and other stakeholders. It then looks at the need for a manufacturing strategy to set goals for manufacturing operations, and identifies the main decision areas of the manufacturing strategy as the mode of production, product variety and flexibility. It describes the infrastructure needed within the company to provide the required manufacturing performance, and identifies CAPM as one element of the infrastructure.

Chapter 2 looks at the setting of competitive goals for the manufacturing company, and identifies the measures of performance which can be used to evaluate the performance of the manufacturing company and of the CAPM system. Definitions of CAPM are presented, together with a description of the principal functions of CAPM systems and some of the common terms in use.

Chapter 3 looks at the control of stock. The kinds of stock that arise are described, as are the reasons for holding stock and the common approaches to the management of stock levels. Some of the assumptions of the common stock-holding policies are challenged.

Chapter 4 describes the family of CAPM systems known as MRP. The modes of operation of the main types of MRP are described, and some of the dangers of the MRP approach are explained.

Chapter 5 describes the production control philosophy known as 'Just-In-Time'. The principles of small batches, elimination of waste and continuous improvement are described, together with the relationship between JIT and 'Total Quality'. The variations of the *kanban* control system are described, and the main arguments against JIT are dealt with.

Chapter 6 looks at the work of Goldratt. It explains his Theory of Constraints and the rules of production management put forward in *The Goal* and his other work. The operation of a scheduling tool based upon this work is described.

Chapter 7 looks at the two main ways of organizing a factory for batch production of discrete items. Two alternative approaches, process organization and product organization, are described and their applicability is considered. The principles of Group Technology, cellular manufacturing, and part families are described.

Chapter 8 follows on closely from Chapter 7 and looks at the techniques which can be used to identify groups to establish a product-focused factory. Various techniques for the identification of part families and machine groups are considered, and their application is discussed. The final section looks at the design of work cells themselves.

Chapter 9 looks at Period Batch Control, possibly the simplest and most powerful production management technique, which is suited to product-focused cellular factories. The operation of PBC is described and the setting of the period length and the production timetable are explained. The role of the cell leader in performing local scheduling tasks is explained, and the possible variations of PBC for special requirements are described.

Chapter 10 considers the question of developing a new CAPM system to meet the needs of a manufacturing company. Systems theory is used to provide a theoretical background to the analysis of systems in organizations, and the concept of the system life cycle is used to describe the various phases of activity in the development and operation of a new system. Each phase in the life cycle is considered in detail. Techniques which may be used in the development of new systems, including the use of $IDEF_0$, are described in the appendices.

Each chapter is followed by a series of questions to help the reader to review understanding of the material covered, and to provide the basis for group discussion work.

The case studies provide the reader with descriptions based upon real-life companies, together with discussion questions. These provide a good basis for discussion work and help the reader to consider some of the problems of CAPM in the complex situations which arise in real factories.

Acknowledgements

The idea for this book came when my two colleagues Gerald de la Pascua and Andie Hallihan completed their Master of Philosophy theses (Hallihan, 1992; de la Pascua, 1992) in the field of production management. It occurred to me then that their descriptions of the basics of CAPM would be of great value as an introduction for later students. With their permission, I began to expand and develop some of their chapters to form the beginnings of this book. Since then, their work has been greatly changed and rewritten, although they will still be able to recognize some parts. I take responsibility for the state it is now in, but those two are to blame for getting me started!

This work would not have been possible without the patience and support of my wife and children, who have hardly seen me for ages. I must also thank my colleagues at the University of Plymouth and elsewhere for their assistance and various people for allowing me to use case studies based upon their operations: Roger Bassett, Alan Cohen, John Gall, Tammi Greswell, David Hughes, Roger Maull, Richard Morgan, Jim O'Brien, Steven Senior, Andi Smart and Adam Weaver.

REFERENCES

de la Pascua, G.P. (1992) *The specification of a computer aided production management system in a capital goods manufacturing company.* M.Phil., University of Plymouth.

Hallihan, A.J. (1992) *A critical evaluation of the application of group technology in a capital goods manufacturer.* M.Phil., University of Plymouth.

Manufacturing and competitiveness

1

The competitiveness of manufacturing depends on many factors. It is the job of the managers of manufacturing companies to control and regulate these factors to allow the company to best meet its competitive goals and maintain an edge in the market-place. These competitive factors include the best use of available technology for new products and for new manufacturing processes, the use of the most effective organizational and management structures (including the best use of human capital through effective training and leadership), and the correct orientation of the business with respect to its customers and the other stakeholders, including competitors, suppliers, shareholders and employees.

1.1 CORPORATE STRATEGY

A manufacturing company need not be a huge corporation to have a corporate strategy. The term 'corporate strategy' is used to describe those choices made by the whole company in the way it does business. In small companies the corporate strategy is indistinguishable from the choices the company makes about its shop floor operations, whereas in the multinational corporation strategy may be made by a corporate staff who might be almost unaware of the company's manufacturing operations. In either case, it is useful to think of corporate strategy as setting the direction of the company, while its manufacturing operations will be used to achieve the required goals. The aims of manufacturing may be to satisfy the customer, but it is the corporate strategy which determines why the company exists and for what reasons the company must trade.

1.1.1 WHY DOES THE MANUFACTURING COMPANY EXIST?

Companies exist for different reasons, and pursue different goals.

- A scientist may establish a company to allow inventions to be brought to market, possibly allowing the development of new ideas so that time can be spent on what is effectively a hobby. Future developments might be to increase the range of projects the enthusiast can become involved in, while providing the opportunity to avoid having a proper job.

- An entrepreneur may set up a manufacturing company to bring together people with various skills in order to make money from a particular market opportunity. When the opportunity is over, the company may be allowed to continue if something else comes up, but in the meantime the maximum amount of profit will be taken, and investments will not be made for the long term.

- An engineer may retire early and invest his pension in setting up a small manufacturing company. As long as the company provides enough income for him to live, he is not interested in developing and growing the business but in having plenty of free time for his family and golf.

- A company may be set up to satisfy a real need in the interests of the customer, such as to supply water pumps for developing countries or to provide aids for the disabled. Profits would not be drawn except to ensure the growth of the company so that its work can increase.

- Many companies exist to provide the maximum return to their shareholders, but this can be in the form of capital growth (the company and its shares being worth more), in the form of the highest possible dividend paid out each year, or a balance between the two.

- For some companies, such as some long-established and well-known names, the corporate goal may be to stay in business, to preserve the company intact and to maintain the personal reputations of the directors. Returns to shareholders may be less important than long-term survival.

Customer satisfaction is often referred to as an aim of manufacturing companies, but in most of the examples above the customers' interests do not appear. Most companies do not exist to serve their customers. For most manufacturing companies, customer satisfaction is necessary in ensuring repeat business and developing a good reputation, but it is not usually the main aim. It is important to know the true aims of the company in order to understand the decisions which are taken and to help the company develop in the required direction.

Hayes and Schmenner (1978) identified four basic elements of corporate strategy which were of particular relevance to manufacturing. They provide four areas in which the company's attitudes may be characterized:

Dominant orientation

A company may be oriented towards a market, to materials or products, or to a particular technology. Market-oriented companies may set out to understand fully and provide for a particular market, such as the market for surfing gear or the market for navigation instruments. Material- or product-oriented companies may be tied to a particular product, such as steel strip, and will develop uses for their material in a variety of markets, such as the car industry, the building industry or the white goods industry.

Pattern of diversification

Will the company widen the scope of its activities into new markets or new technologies, or will it concentrate on a more narrow set of business activities?

Attitude towards growth

Does the company aim to grow, finding new ways to increase its scope, or does the company prefer to focus upon a particular business area and accept the rate of growth which may come from success in that area?

Competitive priorities

Does the company prefer to maintain a high profit margin by selling a premium product in a narrow market, or does it prefer the volume business, where it may have to compete upon price? Does it address its image as a producer of high-quality items or of low-cost items? Does it compete on the basis of reliability and dependability, even though its product may not be the highest performing? Does it compete on the basis of tailoring a product to a customer's individual requirement, or by being able to bring new products to the market quickly?

These questions are not easy for a company to answer. It is generally easier for managers and directors to avoid setting down such ideas, and to take decisions such as whether to develop a new product or to buy a new machine on the basis of intuitive judgement. Such intuition may come from a good degree of shared understanding between the company's top team, developed by working together over a period of time, but if the strategy is set out more clearly it may provide a better basis upon which to do business.

The question of competitive priorities is also addressed by Porter (1985), who categorizes strategies under the headings Cost Competition, Differentiation and Focus. According to Porter, companies must decide whether to compete on the basis of cost or differentiation, and then identify the appropriate focus.

Cost competition provides competitive advantage by carefully controlling costs either to deliver products to the market at a lower selling price, thus gaining volume, or to continue to sell at a price similar to that of competitors while making higher profits through lower material and production costs.

Differentiation allows the company to compete by providing products which have noticeable advantages to buyers, such as improved features or higher reliability. Differentiation may call for continual development and revision of products to continue to provide the best set of features, or it may call for products to be made to specific customer requirements. It may also depend on the package of services which support the product, such as after-sales maintenance and guarantees, or financial arrangements.

Focus is applied to identify the type of products and markets in which the company will operate. A key test of a strategy put forward by Hayes and Schmenner (1978) is to see if the company is clear about the limits of its operations – will it say 'no' to proposals which will move the company into new areas?

The question of focus was first identified by Skinner (1985) who points out that to be really effective the manufacturing business should focus on a 'limited, concise, manageable set of products, technologies, volumes and markets' so that it does not dilute its efforts by trying to do too many different things. The company cannot expect to be successful at increasing market share, maximizing shareholder dividends and developing new products in different areas or technologies all at once. The notion of focus helps the company to decide what it means to achieve, and to coordinate its efforts.

Once it is clear what the aims of the company are, it can begin to determine how manufacturing operations can be used to provide the required performance. It should not be assumed that a manufacturing company will make everything which it has decided to supply to its customers. It may be possible to satisfy the business requirements by simply retailing the products of other companies, together with expertise such as project management or maintenance.

1.2   MANUFACTURING STRATEGY

The need for a manufacturing strategy has been put forward by many writers, notably Hill (1993), Skinner (1985) and Hayes and Wheelwright (1984). A common element of these views is the need to match the performance of the company to its environment. Mendelow (1981) describes the environment within which the company must operate as consisting of the various stakeholders who are affected by the business or who have an interest in the operation of the business. These include shareholders, government, customers, suppliers, lenders, employees, society and competitors. These stakeholders provide an environment within which the company must operate and set the requirements which the company must fulfil.

A manufacturing strategy can be formed to ensure that the company manages its activities in a way that will satisfy the likely demands of customers, employees etc., and which will take into account the likely actions of competitors, suppliers, governments etc. in order to achieve the aims of the business or corporate strategy. The manufacturing strategy may be stated in a policy document approved by the company's senior managers, but it must be continually updated as business requirements change. Many authors now see the process of manufacturing strategy formulation as a key to the successful management of the manufacturing company, and even as more important than strategy itself.

Corporate strategy, as described in the previous section, describes *what* performance is required of the company and *what* approaches it will take to competition and growth.

Manufacturing strategy determines *how* the company will address these requirements by the most appropriate use of the resources available.

1.2.1 THE NEED FOR A STRATEGY

The purpose of a manufacturing strategy is to provide a basis for the coordination of decisions and activities throughout the organization. In order to compete successfully, the company must focus its activities on one particular set of goals, which must be communicated throughout the organization so that everyone pulls in the same direction. If there is no coordination, different parts of the company might work against each other.

Consider a case where strategy is poorly defined. The factory manager wishes to reduce the cost of manufacturing through the increased use of automatic equipment. The sales manager wants to increase sales. In order to increase sales, orders are taken for all kinds of special products with different features for each customer. Competitors are less willing to customize, so this approach keeps the order book full. However, the factory manager has been investing in automatic equipment, which is cheap to run but expensive to reset for each product variant. Thus the factory manager sees the sales force chasing the wrong type of orders, which do not suit the factory, and the sales manager sees the factory buying the wrong type of equipment for the orders coming in. The two managers are both trying to do the best for the business (as each sees it), but the lack of coordination results in excess costs due to unsuitable machinery or orders, depending upon which way you look at it.

At least two strategies are possible. The business could decide to focus upon standard products, which are cheaper to produce, or upon customized products which may command higher prices. This is a strategic decision and cannot be taken at the level of either of the two managers (except by consensus), but must instead be decided by a higher level of management.

This simple example shows a trade-off which should be simple to sort out (although it is based on a real problem which existed in a real company). Clearly, a manufacturing strategy must deal with the type of products to be manufactured and the way orders are to be won in addition to the type of manufacturing equipment to be used. A more comprehensive strategy will deal with all aspects of the business, including personnel, training, factory location, materials, components and the make-it-or-buy-it decision, supplier relationships, quality control, information systems, research and development, and marketing and sales direction.

Unfortunately, many companies have 'strategies' which consist of unhelpful catchphrases such as those shown in Box 1.1 and Box 1.2.

Box 1.1   A strategic orientation?

'In our business, the customer is king.'

A king is a symbol of power and authority. But he is also rather distant and his detailed requirements are not usually well understood. To say that the customer is king is a reminder of the importance of the customer, but it gives no clues as to what actions are required of individuals. If you want to get people to work in a certain way, you'd better tell them what you want.

If the manufacturing business is to compete effectively against other companies, it must not be distracted by competition inside the company. The manufacturing strategy must deal with all these questions, and it must provide the right answers, since the competitiveness of the business as a whole depends upon them.

Box 1.2   A focused factory?

'We design and manufacture anything the customer orders.'

No you don't! There are always limits. What you mean is that you don't want to tie yourself down to any particular strategy. You don't want to tell the rest of the company how you decide to accept or reject orders. But if a strategy was articulated, your colleagues could help configure the business in the way you have in mind.

To establish these answers is not the end of manufacturing strategy. In fact, the activities of competitors and other stakeholders mean that the markets are constantly changing, and manufacturing strategies must also develop to meet, anticipate, or dictate the changes in the market-place. The company must always be aware of the new rules of competition, but it can also change them. Strategy formulation is therefore a vitally important business activity, and strategies must not be seen as fixed but as evolving. Although each company has its own style, which provides its uniqueness and hopefully its reputation, these attributes must always be seen in the light of the current situation.

1.2.2   A MARKET-BASED STRATEGY

The view of strategy taken by Fine and Hax (1985) is that the company operates in a series of markets within which it must compete. A market may be seen here as a collection of competing alternatives within which a choice may be made. Within each market, the value of each alternative is seen against its cost, which may be a function of supply availability and demand for that resource. These markets are:

- *The markets for products and services*
  The company must ensure that it is competitive on the various characteristics of its products or services, such as functionality, price and timeliness.

- *Technology markets*
  Awareness of the full range of available technologies, including new developments and the incorporation of the most advantageous technologies both in products and in manufacturing operations.

- *Factor markets*
  The market for component parts and materials, available through a range of suppliers, which may develop in various ways affecting availability, alternatives and quality.

- *Labour market*
  The market for employees, which may be characterized by the availability and cost of various skills.

- *Capital market*
  The market for investment into the company from a range of sources which have varying requirements, and for which the company both selects and competes.

The strength of this approach is that it encourages the company to anticipate trends in each market and to set targets to meet. For example, in the product market it may see competitors' prices being reduced, and may decide upon actions to reduce its own prices or to win business through some other aspect, such as differentiating its product from the competitors' by increasing functionality or quality, or by addressing a different market area; in the technology market it may see the germination of a new approach which it may decide to exploit, such as the use of composite materials in the car industry. These trends provide the company with an understanding of the way its own activities must be developed.

The idea of markets reminds us that in each of these areas there are competing alternatives. The company must decide which of these alternatives it will address, and it must decide how to go about it. These decisions form the basis of manufacturing strategy.

A respected writer on manufacturing strategy, Hill (1993), also sets manufacturing strategy within the context of corporate strategy. Hill describes the strategy-making process in many companies as an iterative process taking into account corporate strategy, marketing strategies and the way different products win orders against the competition, and which sets requirements to which manufacturing operations must conform. The manufacturing decisions about the most appropriate way to manufacture the products and the supporting infrastructure must then be made to suit these requirements.

Hill argues, however, that for companies to develop sound strategies the manufacturing decisions must be much more closely linked with corporate decisions, particularly those concerned with marketing. The link is established by allowing manufacturing requirements, such as for investment, to be made a part of the iterative strategy-making process, and by considering the way different products win orders against the competition. This forms the common area of interest between manufacturing and marketing responsibilities.

For each product or product family, Hill advises analysis of the *order-winning* and *order-qualifying* criteria.

Order-winning criteria are those which influence the customer's purchasing decision, and may include criteria such as cost or delivery lead time. These are applied when the customer faces a choice between alternative products which are otherwise acceptable.

Order-qualifying criteria are those which allow the product to be considered by the customer. These may be the basic standards of functionality or quality without which the product would not be considered in a particular market or for a particular application.

Different criteria are applied to different products by different types of customer, which means that a good degree of market knowledge is needed to set the requirements for manufacturing. The capabilities of manufacturing may present the company with the opportunity to move into new markets, if manufacturing capabilities are considered in the corporate strategy-making process.

1.2.3   A CAPABILITIES-BASED STRATEGY

While each company defines its markets in a different way, so that it is unusual to find two companies who compete in every aspect of the business, the market-based view of strategy would lead us to suppose that companies operating in similar markets would pursue similar strategies. However, with increasing globalization of markets and increasing competition, market-based strategies fail to provide the competitive advantage which each company seeks. Following the market inevitably means following the competition, rather than generating new advantages for competitors to follow and changing the nature of the market.

A more dynamic approach is required which allows companies to jump from one market position to another, and to compete in new ways. Often this comes about through the introduction of a new product which either creates a new market or which moves the company into a market in which it is different from its competitors. Such a jump comes about not through examination of the company's current market position but through a creative analysis of the company's internal strengths.

Strategies for developing the company's competitiveness should take into consideration the new capabilities which the organization might learn from the change, in addition to the value of the change itself. Hayes and Pisano (1994) cite the case of Hitachi-Seiki, who developed pioneering numerically controlled machine tools which at first were very unreliable and gave little advantage over conventional machines. The company pursued the goal of developing computerized automation through a series of projects which required new skills, and which provided experience of dealing with all the problems which arose. This resulted in the creation of a pool of knowledge about automation which the company could exploit by designing special systems for customers. Besides developing a product, which it might have abandoned if the performance of the product had been the main aim, Hitachi-Seiki had developed capabilities in their staff which gave them a real competitive advantage.

In strategy-making, the company must not limit itself to the current advantages which will be gained by developing a particular product or by adopting a new approach but should also consider the type of experience which might be gained during the development. This may allow the company to learn something which provides a new competitive advantage which may provide the opportunity to compete in new ways, rather than just matching the performance of the competition.

Besides making strategies which are determined from the external market and which lead the company to reconfigure internally (an 'outside-in' strategy-making approach), attention should also be given to the opportunities which may come from an 'inside-out' approach.

1.3 STRATEGIC DECISIONS IN MANUFACTURING STRATEGY

Important decisions which make up the formulation of the manufacturing strategy include the mode of production, the extent of product variety and the need for flexibility.

1.3.1 MODE OF PRODUCTION

The mode of production used in a manufacturing business is often described on a continuum which ranges from the production of commodity materials to the production of one-off unique products.

At one end of the continuum, the volume of production is so high that the products are no longer counted but are measured by physical mass or volume. This includes some products such as oils, chemicals and other raw materials. These products may be classed as commodities, and this form of production is often termed 'mass production'.

At the other extreme, products are produced in very small numbers and are characterized by their uniqueness. At this extreme would be found unique civil engineering and construction projects, some shipbuilding projects, and projects to install production machinery in factories.

Between these two extremes lies what is often thought of as manufacturing industry, in which various companies configure their operations and their strategies to suit the particular volume and mix of products for their business.

Writers such as Browne *et al.* (1988) describe the intervening range of manufacturing situations in terms of the modes of production that may be used in these positions. This is convenient, since the terms high and low volume are relative terms and are not very meaningful. However, the mode of production can be chosen for any particular manufacturing strategy, and is not always a direct function of production volume and variety. Production modes are often described as mass production, flow line production, batch production, jobbing manufacture and individual projects.

Mass production is the production of products such as sugar or petrol which are sold by mass or volume, by plant which runs as a continuous process. These are often regarded as beyond the scope of general production management, since their production is concerned less with the logistics of scheduling materials and machines, as in a factory, than with the regulation of a complex plant such as an oil refinery whose operation depends upon the nature of the process. In this respect mass production has similarities with the extractive industries in the production of coal and chemicals. However, the term 'mass production' is also applied to discrete items which are produced not individually but almost as a continuous process. This includes very high-volume items such as some plastics mouldings and some food, drink and tobacco products.

Flow line production describes the mode of production in which discrete items can be said to 'flow' through a series of locations at which different operations are performed. The locations may be known as a 'production line' or simply a 'line'. This style of manufacture is often used for high-volume items. Benefits of flow line production are based upon the division of the production activities into a series of small tasks which are repeated by each operator for each product. This allows specialist work study engineers to develop the quickest way of performing each activity, and since the flow of work is direct from one operation to the next, no time need be wasted between operations, so the time taken to produce the complete item is at a minimum. However, there may be considerable cost and time penalties in this style of manufacturing when it becomes necessary to change the line from producing one product type or model to the next. Flow line production is commonly used in the manufacture of high-volume items such as motor cars, televisions and 'white goods' or domestic equipment.

Batch production is used when smaller numbers of items of a particular type are required, so that production must change more often from one model or product type to another. This generally means that the factory is not arranged into flow lines. Machines and equipment are commonly grouped by type of machine, by product family or by historical accident. In batch production, batches of parts are moved around the factory from one operation to the next. Since at any time there will be many batches of different items circulating through the factory, and their production routes and times are likely to be different, it is likely that work arriving at a work centre may have to wait until another batch of work is completed. This leads to some delays, but there is great flexibility as batches may be routed around the factory in any way to suit a particular item.

Considerable effort in the production management area is concerned with the difficulties of managing batch production. Some of these are:

- It is difficult to estimate operating time for batches of a new item.
- It is difficult to forecast completion time and total lead time through the factory.
- It is difficult to prevent uneven workload in different areas as the product mix changes.
- It is difficult to maintain the quality of items that are produced for the first time.
- It is difficult to measure performance and to set targets.

Jobbing manufacture is the production of very small numbers of items to specific customer requirements. Production resources tend to be multi-purpose, which means a high level of flexibility but generally slower production. A highly skilled and flexible workforce is likely, and they may develop tools and techniques as required for particular items.

Individual project manufacture is used for individual items of some complexity. Production resources may be brought in as required, such as in civil engineering projects. The product is generally built *in situ* rather than moving around a factory, although some are built and tested in a factory, then disassembled and transported to be installed. Component parts may be produced by jobbing manufacture.

The modes of production tend to reflect the volume of production. For instance, batch production is often used where a factory produces a range of products in medium volumes. However, there may be benefits in producing the same product mix by a flow line if it can be arranged.

The volume of production within the factory does not absolutely determine the mode of production. To an automotive components manufacturer, a volume of fifty items per month would be considered a low-volume item and might be produced in a single batch each month, while the higher volume items might be produced continuously. On the other hand, to a company in the aerospace industry a volume of fifty per month might be the highest volume item, and might be produced in a continuous flow arrangement.

The relationship between the mode of production and the volume of each product is shown in Table 1.1.

Table 1.1 Modes of production and volume

Mode of production	Customer order quantity	Work flow	Production equipment
Mass production	By volume or mass	Process flow through plant	Special plant (may be unique to product)
Flow line	Large number	Along flow line	Special
Batch	Few too many	Many possible routes, some common routes	Some special, some multi-purpose
Jobbing	One-off, or a few	Many possible routes	Multi-purpose
Project	One-off	No movement	Multi-purpose equipment brought in as required

The mode of production has important implications for the selection of the most appropriate production management system. This issue will be considered further in the next chapter.

1.3.2   PRODUCT VARIETY

The variety of products in production at any instant is a good indicator of the complexity of the production management problem.

It is important not to regard the range of product variety as beyond the scope of manufacturing strategy, perhaps being dictated by another part of the organization, such as the marketing department. The effort required to manage the production of a wide variety of products is greatly increased by the extra components and materials that must be procured, the extra drawings, works instructions and manufacturing documentation that must be produced, the increased complexity of scheduling a higher number of operations around the factory, extra tooling requirements etc.

Wortmann (1989) provides a classification of the variety of production. If each product is designed to the customer's specific requirements, then extra effort is required in designing, developing and testing the product. Products in this class are called *Engineer to Order* (ETO). This includes many companies whose products are specialized, of high value and have a long lifetime, such as capital goods manufacturers.

Where designs already exist, products may simply be made to order. This class of product is made up mainly of unique parts which are not used in other products, because of their uniqueness, or which cannot be held in stock because of the very low likelihood of any particular one being required. Products in this class are known as *Make to Order* (MTO).

Many companies attempt to limit the complexity of their product range by using a modular approach which means that a range of standard items are manufactured, while the variety of end products is provided by being able to assemble different standard items into the end product. This class of product is known as *Assemble to Order* (ATO). Products in this class include motor cars and many kinds of industrial products, such as cranes or computers. The advantage of this class of products is that the manufacture of the components can be smoothed out by manufacturing them to a forecast without worrying about obsolescence, since most of the items can be expected to be sold. However, this means that production has to be forecast and a stock of components has to be held to cover the variations in demand from the forecast.

The final class of product is those which are made to a forecast of demand and placed in stock. This has the advantage that customer orders can be satisfied immediately, but it means that inventory has to be tied up in finished goods. The accuracy of the forecast is the key to the economics of such a policy, and this class of production is generally only used for items which are required in high volumes and where variety is low enough to hold stocks of each product. This class of product is called *Make to Stock* (MTS).

The extent of the range of items a manufacturing business produces tends to be inversely related to the volume of production. If there is a high number of different products which a particular company can produce, then each tends to be produced in lower numbers. Conversely, if there is only a very narrow range of products, they tend to be

Table 1.2 Product uniqueness and variety

Lifetime	Consumed – no lifetime	Short lifetime	Frequently replaced	Infrequently replaced	Long lifetime	Long lifetime
Usage	Commodity	Commonly used items, especially components used in other products, consumable items	Common consumer durable items	Varied applications	Special products for specific applications	Major capital investments for unique applications
Volume of production	Continuous	Very high volume	High volume	Low volume	Products made singly or in small number	One product in progress (project)
Product range in business area (variety)	Narrow range	Narrow range	Small range of standard products	Wide range with some repetition	Very wide range of product alternatives	Infinite range – product designed to suit situation
Example product	Sugar, petrol	Fasteners, cutting tools, house bricks	Domestic electrical goods, motor cars	Electrical winch, crane, measuring instruments, musical instruments, furniture	Ship, equipment for power station	Power station, Channel Tunnel

produced in greater volumes. The variety or uniqueness of the product seems also to relate broadly to the product's lifetime and usage, as shown in Table 1.2.

Consider a factory with only very few products. It may decide to use a production line for each item, so that production can run continuously. With more products, it may become impractical to provide a production line for each, as some lines may not have enough volume to justify the machinery. In order to produce them with the same resources, the resources may be time-shared by the use of batches – one batch of each item at a time, so that in turn all can be produced, although some waiting is involved. This is 'batch manufacture'. As variety increases, the demand for each item may be so small that batches often consist of single items. This is known as 'jobbing manufacture'.

Where there is high product variety, towards the ETO and MTO end of the spectrum, production management problems tend to be more concerned with the links between the design process and the manufacturing process, and with the scheduling of work through the factory. At the lower variety end of the spectrum, where ATO and MTS products are found, production management is more concerned with the problems of forecasting sales and determining the correct levels of stock to hold for each item. This spectrum is shown in Fig. 1.1.

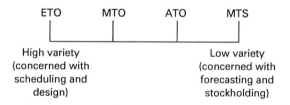

Figure 1.1   The spectrum of product variety.

It is also worthy of note that, at the high-variety end of the spectrum, the manufacturing company is much more closely involved with each customer, since the design has to be formulated to match each individual customer's requirements. At the lower variety end of the spectrum, the customer may have no contact at all with the manufacturing company except to purchase the final product from stock. Correspondingly, at the high-variety end, where the customer is more closely involved, all the activities in the manufacturing company, from contracting and design right through manufacture, are carried out while the customer is waiting. At the lower variety end, time pressure comes from the need to maintain the required stock levels as the market fluctuates, and an individual customer is less likely to wait for an order to be satisfied.

1.3.3   FLEXIBILITY

Many writers on manufacturing in recent years have pointed out the need for companies to become more responsive to rapidly changing markets. New products are being introduced more and more frequently, often embodying the latest possibilities brought about by new technologies.

For example, the manufacturers of home video cameras had to develop their products very quickly to keep up with each other's advances in technology. The state of the art moved from large cameras with shoulder-bag video recorders to 'camcorders', which combined both camera and recorder in one unit through the development of a new tape cassette. Next came further size reduction into a unit little more than the size of a hand – the 'palmcorder' – through the extensive use of surface-mount technology, reducing the size of the electronic circuits. New technologies can lead to obsolescence of existing products, requiring companies to adopt new methods.

Such a summary vastly undervalues the amount of development work necessary to overcome the technical hurdles to create such products, while at the same time improving their ease of use, maintaining extreme reliability and reducing the selling price. Manufacturers have had to invest great effort in the product development process, but the continual development of product ranges is only possible if manufacturing operations can keep up with the necessary pace of change. With each manufacturer offering several products and the range changing completely at least once every six months, the demands made upon the manufacturing function to produce defect-free products to new designs using new equipment are much greater than the demands made in earlier eras, where products could be produced in great volume for years at a time. When competition was less intense, it was common to take the first few weeks or even months to iron out quality problems, but now the need for competitiveness means the cost of a faulty product or a wasted week is enormous. Contracting product life cycles demand more frequent introduction of new products, which require new designs, equipment, materials and methods.

Apart from the need to develop products quickly from concept to full production, increasingly wide product ranges mean that factories often need to switch production from one product to another. (Most factories have too many products to devote separate facilities to each product.) In order to even out the supply of products, batch sizes must be fairly small and changes must be frequent. For example, if a factory produces twelve product types, it could produce the year's expected requirement for product A in January, product B in February and so on, holding stock until needed. This requires the minimum number of changes in the factory. However, it means holding a vast amount of stock, to satisfy an aggregate demand which may call for some of each product every month.

In a fast-changing market, product sales volumes may change suddenly, requiring a changing mix as some products decline and others develop. As obsolescence can be caused by factors beyond the company's control, such as the introduction of a better product by a competitor, such changes are often unpredictable. If stock is held it may become obsolete before it can be sold. The alternative is that the factory should regulate production to match demand, which means producing a number of each product every month, every week or even every day. To do this the factory requires the flexibility to change from one product type to another without incurring lost time.

Some possible solutions

Some companies attempt to resist competitive pressure by identifying a niche market. This is a narrow product area in which there is little or no competition. Some new products, such as a new drug, automatically provide a niche, but only as long as no serious competitor moves into the niche. A market change may mean that the product, despite its uniqueness, is no longer required. This happened to both the Sony Betamax and Phillips 2000 home video recording systems, both of which were regarded as technically superior to the JVC VHS system which became the standard. The niche strategy may leave the company more open to the consequences of market changes, because ways of coping with change, such as developing new products quickly, will not be developed.

Under certain conditions it is feasible for companies to attempt to control the market, through influencing regulations, taking over their competitors or putting them out of business by flooding the market with products at artificially low prices (dumping). All these practices are generally seen as unfair ways of competing and are subject to legal regulations in most countries. On the whole, few companies can use such tactics, and if the company is not generally in a good competitive position the benefits gained by such actions are likely to be overtaken. All manufacturing companies must develop ways of adapting to changing competitive circumstances. Both Sony and Phillips were able to survive the damage caused by VHS.

How flexibility may help

Flexibility is often mentioned as a goal of manufacturing strategy, but as Slack (1987) points out it is not one that has a clear meaning. Flexibility is used here to denote how well a company can respond to changed requirements, whether imposed by the needs of the market (reactive) or by the decisions about innovation which the company takes for itself (proactive). Slack was able to discern four distinct types of flexibility:

Product flexibility: the ability to introduce novel products, or to modify existing ones.

Mix flexibility: the ability to change the... products made within a given time period.

Volume flexibility: the ability to change the level of aggregated output.

Delivery flexibility: the ability to change planned or assumed delivery dates.

Companies which have high flexibility in any of these areas are able to make the changes called for at low cost and in a short time period. A company with high product flexibility can introduce a new product range quickly and cheaply, because it has developed rapid design routines, flexible tooling, standard production methods etc. Companies which are less flexible will find that to make these changes is slow and costly. A company with low delivery flexibility will find that to change delivery dates causes expensive rescheduling problems such as overtime working, stock shortages, and delays and other expenses caused by disruption. A more flexible company would have the resources to work around these changes.

The terms 'response' and 'range' were coined by Slack to provide another way of describing flexibility. Response relates to the company's ability to change in terms of the time and cost involved. Range relates to the set of possible outcomes from a particular type of change. For example, the flexibility for a company to change its production volumes (while still operating profitably) may be very high within certain limits, but changes beyond these limits would require expensive retooling, retraining or some other change which may be expensive. Thus the company's volume flexibility may be high within a certain range. Similarly, product flexibility may be very high as long as the company is only required to develop new products within a certain range of technology or functionality, but much lower if the change involves something unfamiliar.

Much attention has been placed upon the flexibility of machine tools, especially with the development of Flexible Manufacturing Systems (FMS) – machines which have a very fast response to change within a given range. However, it appears that flexible technology on the shop floor has to be supported by a flexible infrastructure (Meredith, 1986; Maull *et al.*, 1990).

1.4 THE MANUFACTURING INFRASTRUCTURE

(This section is adapted from Childe (1991a,b).)

The infrastructure consists of the systems which support the manufacturing activity. It has been defined by Hill (1993) as the 'structures, controls, procedures, communications and other systems... the attitudes, experience and skills of the people' and by Skinner (1985) as the 'policies, procedures and organization by which manufacturing accomplishes its work, specifically production and inventory control systems, cost and quality control systems, work force management policies and organizational structure'.

Meredith (1986) also describes the infrastructure as 'the network of non-physical support systems that enable the technical structure to operate'.

The infrastructure is critical to the manufacturing operation, since manufacturing can only respond to the market-place at the rate at which market information is translated into instructions for manufacturing in the form of orders, schedules etc., and at the rate at which new designs, equipment and materials can be provided.

For instance, production may be held back by the time taken for designs to be modified, for orders to be processed, for machines to be repaired or for personnel to be trained. These activities are often out of the control of the Production Manager, and yet they affect production.

Slack's four types of flexibility cannot be provided simply by making available flexible shop floor machines and equipment. The flexibility of the infrastructure would restrict even the most flexible manufacturing hardware by failing to provide the support required.

A more detailed understanding of the functions of the infrastructure is required. The provision of orders, schedules, designs, equipment and materials has been mentioned earlier. It is proposed that the functions of the infrastructure are to provide the following inputs, which completely support and control the manufacturing activity:

1. designs
2. materials

3. methods information, such as manufacturing layouts, works instructions, process instructions, quality procedures, numerically controlled machine part programs
4. orders, such as work-to lists, job schedules, picking lists, batch cards and travellers
5. facilities, such as the plant and equipment, machinery, fixtures, tools and gauges
6. trained and motivated personnel.

Thus the infrastructure provides three physical requirements (materials, facilities and personnel) and three informational requirements (designs, methods and orders).

A schematic diagram showing the role of the infrastructure in the manufacturing system is shown in Fig. 1.2.

Manufacturing system

Infrastructure		Physical system
Supply designs	Designs →	
Supply materials	Materials →	
Supply methods	Methods and layouts →	
Supply orders	Orders, schedules, work-to lists →	Manufacture products
Supply and maintain equipment	Maintained plant and equipment →	
Recruit, train, remunerate and manage personnel	Trained and motivated personnel →	

Products →

Figure 1.2 The infrastructure in the manufacturing system.

These six functions of the infrastructure provide a framework which allows the analysis and design of infrastructures in specific company

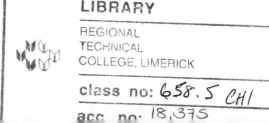

situations. They provide a basis upon which decisions can be taken with regard to the amount of flexibility required in each area.

In this book we are particularly concerned with computer aided production management (CAPM). In this infrastructure model, CAPM can be seen as including the management activities which control the flow of orders, which regulates the production of products, together with the management activities which control the supply of materials. These both involve making orders: for materials from suppliers and for manufacturing operations in the factory.

1.4.1   INFRASTRUCTURE DESIGN

The design of manufacturing infrastructures must be based upon establishing performance parameters for the various infrastructure elements. In any area there will be limits of range and response which will limit the infrastructure's performance. The range and response criteria must be established to allow the infrastructure to operate within a range of situations which can be regarded as likely situations in the business. Thus, as in structural terms a milk bottle manufacturer would not provide sufficient flexibility to switch production from milk bottles to printed circuit boards (PCBs), in the infrastructure area a PCB manufacturer would not train designers in the area of milk bottle design. Although a wide range and a speedy response would seem ideal, the practical answer is to set limits to the desired performance, within which any change could be met. The setting of these limits is thus a strategic question for the direction of the business.

1.4.2   INFRASTRUCTURE FUNCTIONS

The following examples show how flexibility is provided in different areas of the infrastructure to suit different businesses. These are drawn from companies in the electronics industry.

Design

Flexibility of design capability relates to the rate at which new designs can be provided. Company A, a manufacturer of domestic electronic goods, found that increased flexibility and Just-In-Time manufacturing on the shop floor, together with a continuous improvement policy, created a demand for thousands of design modifications per year. The existing procedure became overloaded, and the company began considering ways of improving this function.

Company B, a computer manufacturer, also found that increasing numbers of design changes were required, and was able to implement a computer-based design change management system to facilitate the passing of information between departments and to monitor the progress of all change requests.

Material

Materials can be a severe limiting factor to flexibility. Company B finds that long procurement times, such as three months to purchase an ASIC chip from Japan, are a limit to the company's responsiveness. This has also caused quality and inventory problems due to boards being assembled as components become available rather than in line with the customer order schedule.

In the electronics industry the total manufacturing lead times and therefore the responsiveness to customer demands depend to a large extent on the procurement lead time, since the actual manufacturing lead time is relatively short. In Company C, a manufacturer of defence-related products, steps are being taken to improve the time taken to issue purchase orders and to improve supplier relationships in order to become more responsive.

Company D, a components manufacturer, has improved its ordering process so that all purchasing is managed by one administrator and a typist, who also do other work. This appears to provide the appropriate level of flexibility for this business.

Methods

Company E, a subcontract assembler, has no area with specific responsibility for method engineering, since the business is built around one specific method only and all products are processed through the same route. This company has decided to have no flexibility of methods. If this technology becomes obsolete, the company will face difficulties in responding quickly.

Company F, a component manufacturer, requires little methods flexibility to meet day-to-day changes, which tend to be of mix only. However, it has assembled a highly skilled team for the development of new process equipment to create a specific market advantage.

In some companies, for example company D, methods work is carried out by the design department.

Order processing

Most of the companies investigated are conscious of the importance of processing orders quickly. For example, companies D and F both allow major customers to enter orders by data transfer direct into the orders database.

Company G manufactures tailor-made electronic equipment. Customers ordering from this company are able to speak directly to a design engineer, who can quickly tailor the design as required, provide a price and immediately pass the order to manufacturing.

Facilities

Most of the companies investigated had made little attempt to develop flexibility in the area of providing tools and equipment, generally relying on external equipment manufacturers, tool manufacturers and maintenance contractors. While some were able to design their own tooling requirements, these were provided by external companies. Using external companies was generally seen to be a good solution, and steps had been taken to develop good working relationships with these companies.

Personnel

Flexible personnel approaches have been implemented by almost all the companies. This generally involves operators moving between different operations.

Company F has attempted to de-skill all direct operations and to employ operators on short-term contracts to provide the flexibility to increase and decrease numbers as required.

Company E contracts out some operations and employs outworkers on a subcontract basis.

1.4.3 META-LEVEL FLEXIBILITY

A further aspect of flexibility which may be observed is the flexibility of the company to change the configuration of the infrastructure itself. As we have seen, companies are able to determine the correct level of flexibility for the changes demanded in their businesses as they stand at present, but on-going changes may mean that different kinds of flexibility will be needed in the future. The infrastructure may need to develop flexibility in different areas from what is required now. The ability of the infrastructure to adapt to suit new conditions may be regarded as a meta-level flexibility. Such flexibility is likely to depend upon the skills, attitudes and capabilities of the company concerned.

1.4.4 CAPM AND THE INFRASTRUCTURE

Production management systems bridge two of the six infrastructure areas identified above. These are Orders and Materials.

Production management requires the regulation or control of manufacturing activities in order to meet customer demands. It must therefore issue orders to control what is to be made and when. Thus a principal function of the CAPM system is to derive orders for production based upon customer requirements. As manufacturing activities are dynamic, the CAPM system must also take account of the current status of work on the shop floor, in case late work affects the resources available for current work.

Production management systems often also generate the orders the company issues to suppliers of raw materials and purchased component parts. This arises because purchasing depends upon what is to be used to fulfil customer requirements, whether purchasing items for a specific order or to replenish stocks which will be depleted by an order. It is often convenient to calculate purchasing requirements at the same time as manufacturing requirements, because both depend upon the list of items needed to make a product.

Other functions performed by production management systems may include the collection of performance data relating to manufacturing, such as levels of scrap and re-work, amount of late deliveries to customers, operating costs, productivity and efficiency measures, supplier performance etc. These measurements can be used to help identify problems and target improvement efforts.

The design of the production management system must take into account the nature of the business. Apart from at the most abstract level of description, production management systems, like manufacturing businesses, are unique. The design of the system will depend upon a wide range of factors, including product volume, range and variety; rate of introduction of new products; mode of production; number of workstations; complexity of product routing; complexity of product structure; number of purchased items; number of suppliers; number and type of customer; length of manufacturing and purchasing lead times; and stock-holding policy. All these factors are either direct or indirect results of strategic choice within the manufacturing business. It is therefore essential to make strategic decisions in order to design an appropriate production management system. Since the performance of the manufacturing system as a whole depends on the performance of the other functions of the infrastructure, these must be designed as an integrated whole, according to the current manufacturing strategy.

SUMMARY

This chapter has outlined the importance of a manufacturing strategy for coordination of manufacturing operations for competitiveness. Strategies may be developed on the basis of the state of the markets in which the company chooses to operate, on the basis of the development of the company's capabilities, or from a combination of both. A key element of the manufacturing strategy is the mode of production used. The choice of production mode depends to a large extent upon the volume and variety of products to be manufactured, but is not dictated by these factors. Manufacturing companies increasingly face changes in technology which increase the product range and increase the need to switch from one product to another. Thus flexibility has become an important feature of manufacturing systems. The flexibility of the manufacturing system as a whole depends not only upon the flexibility of the factory and its resources, but also upon the flexibility of the manufacturing infrastructure which supports manufacturing operations. Production management systems form part of the infrastructure in providing the orders to control manufacturing activities to ensure that the right work is produced, and in providing the materials needed for production.

QUESTIONS FOR DISCUSSION

1. For what reasons may a manufacturing company exist? Whose interests may it serve?
2. What benefits does each type of stakeholder gain from a successful manufacturing business?
3. In what ways do stakeholders affect the operation of the manufacturing business?
4. How does a market-based strategy help a manufacturing company to compete? What kinds of business may benefit most from such a strategy?
5. How does a capabilities-based strategy help a manufacturing company to compete? What kinds of business may benefit most from such a strategy?
6. How may a manufacturing company decide upon the mode of production to use? What factors would affect this decision?
7. Think through the classes of product: ETO, MTO, ATO and MTS. Which is most suited to changing demands, and which to orders which are needed quickly? To which class do the following products belong: an electronic calculator, a central heating boiler, a railway locomotive, a car exhaust, an aero engine?

8. What is 'flexibility'? In what circumstances is flexibility important? What factors affect the flexibility of a manufacturing company?
9. What is the function of the manufacturing infrastructure?

REFERENCES

Browne, J., Harhen, J. and Shivnan, J. (1988) *Production Management Systems – A CIM Perspective*, Addison-Wesley, Wokingham.

Childe, S.J. (1991a) Flexibility through the design of manufacturing infrastructures, in *Integration and Management of Technology for Manufacturing* (eds E.H. Robson, H.M. Ryan and D. Wilcock), IEE Management of Technology Series 11, Peter Peregrinus, London.

Childe, S.J. (1991b) *The Design and Implementation of Manufacturing Infrastructures*, University of Plymouth.

Fine, C.H. and Hax, A.C. (1985) Manufacturing strategy: a methodology and an illustration, *Interfaces*, **15**(6).

Hayes, R.H. and Pisano, G.P. (1994) Beyond world-class: the new manufacturing strategy, *Harvard Business Review*, Jan–Feb, 77–86.

Hayes, R.H. and Schmenner, R.W. (1978) How should you organize manufacturing? *Harvard Business Review*, Jan–Feb.

Hayes, R.H. and Wheelwright, S.C. (1984) *Restoring Our Competitive Edge – competing through manufacturing*, John Wiley & Sons, New York.

Hill, T. (1993) *Manufacturing Strategy*, 2nd edn, Macmillan, London.

Maull, R.S., Hughes, D.R., Childe, S.J., Weston, N., Smith, J.S. and Tranfield, D.R. (1990) A methodology for the design and implementation of resilient CAPM systems, *International Journal of Operations and Production Management*, **10**(9), 27–36.

Mendelow, A.L. (1981) Environmental scanning – the impact of the stakeholder concept, *Proceedings of the 2nd International Conference on Information Systems*, Cambridge MA, USA, December 7–9.

Meredith, J. (1986) Automation strategy must give careful attention to the firm's 'infrastructure', *Industrial Engineering*, **18**, 68–73.

Porter, M.E. (1985) *Competitive Advantage: Creating and Sustaining Superior Performance*, The Free Press, New York.

Skinner, W. (1985) *Manufacturing – The Formidable Competitive Weapon*, Wiley, New York.

Slack, N. (1987) The flexibility of manufacturing systems, *International Journal of Operations and Production Management*, **7**, 35–45.

Wortmann, J.C. (1989) Towards an integrated theory for design, production and production management of complex, one of a kind products in the factory of the future, in *ESPRIT 89, Proceedings of the 6th Annual ESPRIT Conference, Commission of the European Communities, Brussels*. Kluwer Academic, Dordrecht, pp. 1089–99.

Production management and CAPM

2

From the previous chapter we have seen that the success of a manufacturing company depends upon the performance of its manufacturing resources. The factory's performance is constrained by the performance of the manufacturing infrastructure, which is itself designed in order to suit the company's strategic decisions about what kind of product to produce and how to configure production resources.

Production management is concerned with taking the company's strategic requirements in terms of its key competitive objectives, whether they be low cost, short lead time etc., and providing the factory with a timetable of orders and the material required to allow these key objectives to be met. The decisions taken at these three levels are shown in Table 2.1.

2.1 COMPETITIVE OBJECTIVES

While the manufacturing strategy sets the context within which production management operates, the way in which production is managed in support of this strategy can still have an enormous effect upon business competitiveness. Good management of production may be the most critical element of success in a manufacturing company, since the management of production has a direct influence upon the satisfaction of the customer, through timely delivery, and production is often responsible for the largest share of the company's resources, and therefore its costs.

Clearly, a major function of production management is the successful completion of customer orders on time and at the minimum cost. Unfortunately, there is a tension between these two objectives. If all customer orders are to be completed on time, one approach would be to keep large stocks of finished products. There would be a considerable cost associated with such a policy, since the stocks would have to be stored and may become obsolete before being sold, and the amount of money expended in purchasing the materials and paying for the manufacture of these items may be lost. In any case, such a high investment would be wrong from a business point of view, since the owners or shareholders could obtain a higher return from investing in a company with a more economical policy. In order to manage this trade-off, it is important to consider the financial aims of any manufacturing company.

Table 2.1 Levels of decision in a manufacturing business

Level of decision	Type of decision
Business strategy	Type of business Legal and financial status Location Research and development
Manufacturing strategy	Product range Selection of key objectives such as cost, lead time, delivery reliability, product functionality and features Extent of product customization Mode of production Make or buy decision Personnel policies Supplier relationships Quality assurance Information systems
Production management	Purchase of material Allocation of material to orders Allocation of production time and resources to orders Achievement of planned output

2.1.1 MEASURING THE PERFORMANCE OF A MANUFACTURING COMPANY

Goldratt and Cox (1984) suggest that the goal of all manufacturing companies is to make money. The overall performance of the business can be measured using the three measures net profit, return on investment and cash flow.

Net profit (NP) is the money left over after the total operating costs have been subtracted from total revenue from sales (turnover).

Return on investment (ROI) is a measure of the benefit gained from making an investment in the company. If an investor considers making an investment in the company, the decision will be affected by other investment opportunities available, where the same investment may secure a larger profit. The total money invested in the company indicates the total value of all the company's buildings, machinery and equipment, stocks of finished and part-finished goods, and raw materials. The return is the amount of profit made, which can be used to reinvest in the business and to pay dividends to owners and shareholders. Return on investment is the ratio of net profit to investment, and is usually to be maximized. A good return is essential to maintain the support of the shareholders. Decisions using the ROI figure include whether to invest in the company and also whether to alter the way money is invested in the company. It might be more beneficial to make a small reduction in profit in one area of the business if this frees up investment capital which can be used in another area to produce a larger return. An example of this might be freeing several million pounds in inventory costs, which could then be invested in a new machine tool or the development of a new product which would provide a higher return overall.

Cash flow is described by Fox (1982) as the 'red line of survival'. Even when a company is making good profits it can still be made bankrupt by failing cash flow. When this happens the company has no money available to pay its operating costs, such as its employees' salaries, or to buy new raw material. Cash flow problems can be caused by large orders which would otherwise be profitable. The company's available cash may be used up before payment for the order is received, thus preventing it from paying regular expenses. If the company is otherwise healthy, it may be able to borrow to cover a cash flow problem, but this leads to interest charges which are an unnecessary cost and reduce profits.

These financial measures are rather distant from the company's actual shop floor manufacturing operations, but the reader will notice that for production management they indicate that costs should be less than turnover, investment should be kept low to provide a good rate of return, cash flow should be smoothed out if possible and the manufacturing cycle time between order and payment should be reduced. These business level indicators can be translated into more specific measures for the production manager. These are throughput, inventory and operating expense (Goldratt and Cox 1984).

Throughput (T) is defined as 'the amount of cash generated by sales'. This is a measure of the value added by the manufacturing company, and is calculated by subtracting the cost of materials used in making a product from the income generated by the sale of the item. By this definition work can only be counted as throughput when it has been sold to generate income. Therefore making larger batches does not increase throughput if some of the batch goes into finished goods storage. The additional products have not been sold, so they have generated no 'cash through sales'. Increased throughput can be expected to increase both the income from sales and the costs of production (since more work has to be done). If the company is operating profitably, the costs rise less than the income.

Inventory (I) is the net value of all the assets, including work in progress, raw materials, finished parts, and machines or buildings. It is very important to note that, unlike standard accounting practice, Goldratt's method of measuring inventory includes all the assets, including work in progress, machine tools and buildings. This makes sense because it emphasizes return on investment. Buying new machines increases inventory and can only be justified if there is an increase in throughput or a reduction in operational expense large enough to compensate for the increase in inventory. An inventory holding cost is sometimes used to quantify the cost of holding an item of a certain value, and is usually reckoned against the bank lending rate. If this rate is, say, 7% then to hold an item worth £100 for a year costs £7 plus the costs of the space to keep it in and the cost of the time and effort needed to find it again. This approach underestimates holding costs, since the business itself must provide a greater return than the bank, so the real cost is the cost of not investing in an item which can immediately be used and sold, such as a piece of material.

A bigger company would be expected to have a higher level of inventory than a smaller one, so the absolute level of inventory is only meaningful inside the company concerned. A proportional measure is sometimes used to compare the level of inventory to the throughput of the company. This is the *stock turn ratio* (STR) or *inventory turns*, which is expressed as the total throughput for a year divided by the average value of stock held during the year. The ratio shows how many times the stock is replaced in the year. Typical values for the STR are 1–5 in industries where products are of high value and long lead time, such as in the aerospace industry, and 5–20 for many batch manufacturing companies. A value of over 100 has been claimed by some companies. These measurements have to be treated carefully. It is usual to calculate the STR on the basis of materials and goods rather than including the value of plant and equipment. It may also be calculated on the basis of total revenue rather than throughput. These differences both increase the ratio.

Operational expense (OE) includes all the costs incurred in operating the business, including wages, maintenance, inventory holding costs, rents/rates and service charges. For a given level of throughput, these expenses should be kept as low as possible.

The relationships between the financial measures and the manufacturing measures are thus:

$$\text{Net Profit} = \text{Throughput} - \text{Operational Expense}$$

$$\text{Return On Investment} = \frac{(\text{Throughput} - \text{Operational Expense})}{\text{Inventory}}$$

2.2   THE GOAL

In order to satisfy the goal of making money now and in the future, the company must aim to increase simultaneously net profit, return on investment and cash flow (Goldratt and Cox, 1984).

Goldratt extends his arguments above regarding the main business indicators and concludes that there are only three ways to increase profits, as described below.

2.2.1   REDUCE INVENTORY

After an initial reduction to a low work in progress (WIP) environment it is difficult to make a significant difference to profits solely by reducing WIP. Because, in Goldratt's definition, the company assets (such as machine tools and buildings) are counted under this heading, the scope for reduction is limited. Reducing inventory by selling excess factory or storage space may be counter-productive, because it may make expansion more difficult.

Production management can address this item by various means which minimize the amount of stock and work in progress, but large gains can only be made in this area if the starting position is particularly bad. Inventory can normally only reach zero when no manufacturing is being done. (An exception to this occurs if it can be arranged for the customer to pay for the final product before the material is purchased. Then the work in progress appears to have no investment value because it has already been sold. However, to assume that the work has no value would imply that it could be left lying around or even scrapped.)

2.2.2 REDUCE OPERATING EXPENSE

Many of a company's operating expenses are fixed. For example, the wages are usually unchanged regardless of whether the workforce is fully utilized or not; it is difficult to make savings on costs such as the costs of buildings and land unless a very large change in the business is made. In any case, these costs are often small compared with the costs of materials and tooling which are expended in production. Some companies try to reduce operational costs by adopting a 'hire and fire' policy, which allows reduction in employment costs at times of low demand. In general, operating expenses may be minimized by reducing waste of all kinds. Production management can address the elimination of waste through making improvements in the way work is done, so that time is not wasted, material is not scrapped etc.

Operating costs, like inventory, cannot be reduced to zero if the business is still to operate. The scope for improvement in this area also depends upon the company being in a bad state to start with.

2.2.3 INCREASE THROUGHPUT

As stated earlier, throughput is the amount of cash generated by sales. Unlike the other two ways of increasing profit, there is no limit to the amount by which throughput can be increased.

Throughput can be maintained and increased in two main ways. Firstly, the production management system must ensure that the work will be completed by its requirement date, so that no sales are lost, and secondly it should process work in the most effective manner so that the maximum volume of throughput can be achieved from any given level of resources.

2.2.4 OBJECTIVES FOR PRODUCTION MANAGEMENT

Among the three ways of increasing profit, the one with the greatest long-term potential is to increase throughput. This is because the other two options are limited in the amount they can be reduced by in the long term.

Throughput can usually be increased to meet market demands, at which point the volume of orders being generated for sales becomes the constraint which limits the profitability of the company.

The most desirable situation is to have the greatest possible throughput with the minimum operating expense and inventory. This should be the objective of production management.

2.3   DEFINITIONS OF CAPM

Computer aided production management (CAPM) is the use of computers to improve the efficiency and effectiveness of production management. It is concerned with what to manufacture, in which order and by when.

CAPM is a broad heading which comprises a wide range of tasks some of which are carried out in the heads of shop floor supervisors and foremen, some by computers and some, traditionally, on the back of a cigarette packet. In some companies the involvement of computers may be zero (although it is now unusual to find a company with no computers at all), while some companies operate sophisticated and complex suites of computer software on mainframe computers or distributed networks, some of which may be fully integrated with other parts of the business such as design engineering, sales and accounting, and sometimes connected through external links to customers and suppliers. Several writers have tried to define the activities which are carried out in CAPM.

Corke (1985) defines CAPM as being:

concerned with the executing of customers' orders, efficiently and economically. It is concerned with

1   knowing at all times what delivery dates can be offered realistically, taking account of existing commitments.
2   planning future capacity to meet sales opportunities.
3   ensuring the right materials are ordered.
4   ensuring that work in progress proceeds through the manufacturing stages in the right sequence.
5   providing flexibility to meet changing customer requirements or priorities without incurring excess inventory.

Waterlow and Monniot (1986) concluded that CAPM should be considered as all the aids supplied to the production manager. They defined the three main areas as:

a) Specification – ensuring the manufacturing task has been defined and instructions produced.
b) Planning and control – planning the timetable, adjusting resources and priorities and controlling production activity.
c) Recording and reporting – recording and reporting production status and performance for liaison with other departments, and future use in specification, planning and costing.

CAPM systems also have the potential to help production managers with their information processing in three main ways:

a) Transaction processing – maintaining, updating, making available specifications, instructions and production records.
b) Management information – providing information for exercising judgements about the use of resources and customer priorities.
c) Automated decision making – deriving production decisions using algorithms.

For the purpose of the recent CAPM Initiative (ACME, 1991), CAPM was regarded as:

The use of computer-based information to support production management functions and to coordinate flows of orders, materials and finished goods.

The author considers CAPM to be a system which includes all the activities concerned with the provision of materials, the regulation of production activities, and the control of stock levels to ensure that customer orders are satisfied. The concept of a system is explored further in Chapter 10.

2.4 THE MANAGEMENT OF TIME

CAPM systems must purchase materials which arrive at the right time, they must deliver products to customers at the right time and they must carefully manage the time of the people and machines in the factory. Before looking at the common features of CAPM systems, it is important to understand the terminology used to deal with time.

The amount of work or *load* on any resource in any given period can be calculated from appropriate measures of time and the details of the work required in terms of the number of batches, components or operations. This can be compared with the amount of time available at the resources concerned, which is termed the *capacity*. Different measures of time are available at different levels of detail which are either based on previous experience of performing the same or similar activities or upon calculations based upon the times for the elements which make up the total activity. Commonly used measures are shown in Fig. 2.1.

Total product lead time may be quoted to customers, and is a guide to the length of time taken to produce a completed product. This may be based upon experience or calculated on the basis of *component and assembly lead times, machine lead times* or *production cell or line periods,* and may include the time taken to process the order and to obtain materials and/or to perform design work. It may also include the time taken for operations which are performed by subcontractors.

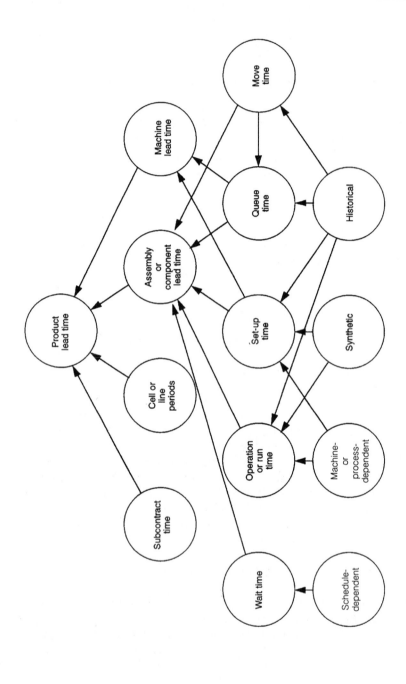

Figure 2.1 Measures of time in manufacturing.

Component and assembly lead times are estimates of the total elapsed time required to manufacture a component or to complete an assembly. These are either based upon experience or calculated on the basis of *operation or run times, set-up times, queue times,* and *move times.*

Machine lead times are sometimes used when it is possible to estimate the lead time for any operation on a particular machine. This may arise when the machine requires long *set-up times* or has a long queue of work waiting, which means the *operation time* for the particular item becomes insignificant. It may also arise when the operation time is not insignificant but is the same for all items, such as in heat treatment. In either case the lead time can be estimated from a combination of the *operation time, set-up time* and the *queue time.*

Production cell or line periods describe the time taken for a component or product to pass through a cell or along a production line. (The concept of a production cell is described in Chapter 7.) These times can be estimated accurately. Production cells are often allocated a fixed time period on the basis of being able to complete all the period's work by the end of the period. Thus the actual variations in time taken are absorbed by the cell and to the outside observer the cell becomes completely predictable (see Chapter 7). Production lines often work at a fixed rate, so the time taken for a product to pass through can be predicted accurately.

Operation times or *run times* are estimates of the time required to perform a particular machining, process or assembly operation. These times may be arrived at from experience, from estimates based upon *synthetic times,* or from the operation being time-dependent (such as many heat treatment processes). Operation or run times include the time taken to load components into machines and to unload them when finished, in addition to the time during which the item is being worked on. In the volume manufacture of small items, times are often recorded per 100 items, to make the numbers easier to handle. Where a batch of items are processed simultaneously, such as in heat treatment, the run time does not depend upon the number in the batch, so the time taken for the process may be recorded as a *set-up time.* In this case, the run time may be recorded as zero, or it may represent the workpiece loading and unloading time, which does depend upon the number in the batch.

Set-up times include the time taken to set up a machine or piece of equipment to perform an operation, plus the time taken to break down the machine ready for the next set-up. Set-up time does not include the time taken to load or remove a workpiece, and is thus independent of the number of items to be processed. Set-up times are normally fixed for a particular work centre, although some production management systems recognize that set-up time can be saved by sequencing work so that jobs requiring similar set-ups follow on from each other. These are called 'sequence dependent' or 'dependent' set-ups. Set-up times are usually determined by experience, although *synthetic times* can be used. Since set-up times are independent of the number of components to be processed at one set-up, they are often the only times recorded for process operations such as heat treatment or plating, where the time taken depends on the process rather than on the number of items to be dealt with (as long as the number is within the capacity of the oven or tank).

Queue times are approximations used to help estimate lead times in batch and jobbing factories where components follow different routings and may often have to wait before being processed by any particular work centre. Queuing times increase greatly as the workload of the work centre increases, so they can only be regarded as a very approximate indication. In many cases queue times also include an element of *move time*, especially when an overall estimate is used between all operations.

Move times are used with queue times in batch and jobbing factories where the transportation of work between work areas takes an appreciable length of time. This may be because of the need to use special equipment to move the work, such as special trolleys or cranes for large items, or because work movement is assigned to a particular group of workers. In either case the move time often consists mainly of the time spent waiting to be moved. It therefore varies with the workload of the work moving equipment or personnel, and is difficult to predict. Move time is often incorporated in queue time.

Wait time is sometimes used to describe the time when a finished component part waits for another item before it can be used in assembly.

Synthetic times are sometimes used to calculate expected or allowable times for operations or set-ups. They provide standard data for elements of work, such as the time taken to reach for a tool from a rack, or to make a solder joint. The time taken for elements of work which depend upon equipment can be estimated by calculations based upon the speed of the equipment, such as the time taken to fasten a screw using an electric or pneumatic screwdriver of a certain speed, or the time taken to manufacture a component in a machining centre based upon the speeds and feeds involved, and the manufacturer's data concerning tool changes and rapid traverse movements etc.

2.5 SOME ELEMENTS OF CAPM SYSTEMS

2.5.1 BILL OF MATERIALS (BoM)

The bill of materials is a list of components required to assemble a finished product. It is used in calculating the requirements for subassemblies and components from the known demand for the finished goods. The BoM can be considered as a 'recipe' for the finished product; in this case the components and subassemblies can be considered as 'ingredients'. Knowing the requirements for finished goods and the 'recipe' for those finished goods means that the requirements for ingredients can be calculated.

2.5.2 MASTER PRODUCTION SCHEDULE (MPS)

The master production schedule contains the requirements for finished goods and dates for their completion. In most companies this is usually a mixture of firm orders and sales forecasts.

2.5.3 ROUGH-CUT CAPACITY PLANNING (RCCP)

Once a master production schedule has been produced, it may be checked by the use of rough-cut capacity planning. Using simplified profiles of each product, showing their work requirement over their manufacturing lead time, the amount of work in any particular resource area in a particular period can be roughly calculated. If the MPS generates loads which fluctuate greatly from period to period, or if some periods appear to be overloaded, it may be possible to reschedule to try to even out the load. This is known as load smoothing. Load profiles, which are a very simplified estimate of workload based on a general product type rather than the exact item, allow RCCP to be done quickly as a simple aid to master production scheduling. Detailed schedules are not made at this stage, since to calculate the exact work requirements on every component is very time-consuming and various iterations may need to be made.

2.5.4 SCHEDULING

Scheduling is the planning of production activities. It may use average or estimated lead times or a combination of run, set-up, queue and move times, which may themselves be estimated or calculated as described above, according to the level of data available and the amount of detail required. There are two forms of scheduling: forward scheduling and backward scheduling.

Forward scheduling starts at the current date or time and adds the time for each activity required in the manufacture of an item. This theoretically gives the earliest time that the item can be completed.

The difficulty with forward scheduling is that it will always give the earliest production date for all parts. If parts are required in sets, such as for assembly or to make up a multi-item customer order, this is not desirable because it means that some parts will be completed before others and will have to wait. Across a whole factory there may be a considerable amount of work which has been made in advance of requirements, which is excess inventory.

Forward scheduling is useful for making delivery commitments for individual components, for example spare parts, which are required as soon as possible.

Backward scheduling starts at the requirements date. From the requirements date the system schedules backwards the time required for each activity. This type of scheduling should give the dates on which the activities should be started in order to be completed on time.

The main problem with backward scheduling is that sometimes the start date is before the current date, meaning that activities are late before they begin. The required product will not be ready on time unless activities can be carried out faster than is normally expected.

Similarly, if a new order has to be fitted into a schedule which contains no slack, there is no option but to delay all other work using the same resources, which will then inevitably be late. These problems are often dealt with by adding overtime working (which increases costs) or by using lead time estimates which contain plenty of extra time for safety. This means all lead times are longer than needed and work in progress is higher than needed.

Combined forward and backward scheduling

The completion date for a product depends upon the assembly time plus the longest component lead time, as shown in Fig. 2.2(a). One way of using both forward and backward scheduling is to base the schedule on the longest lead time item. This part can be forward scheduled to give the earliest possible completion date, and then the rest of the parts can be backward scheduled, as shown in Fig. 2.2(b). This leaves all parts as late as possible, so the work in progress level is low. However, the plan is not robust, as any delay will delay assembly. A more reliable solution is to add a small amount of safety lead time between operations, as shown in Fig. 2.2(c). This has the effect of raising the level of work in progress, while providing more resilience to cope with the delay of any operation. The trade-off between the increased reliability and the increased cost must be determined against the competitive priorities of the company. In many companies this trade-off is handled clumsily by adding queue time to all operations when scheduling.

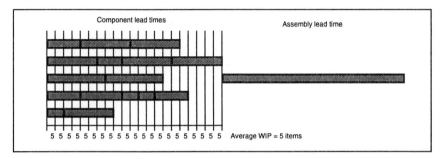

Figure 2.2a   Forward scheduling all items.

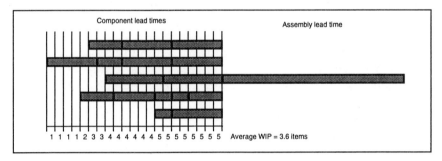

Figure 2.2b   Forward and backward scheduling (no slack).

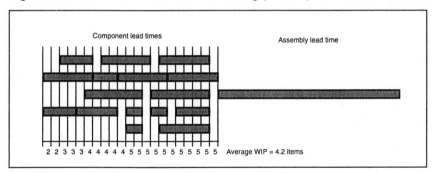

Figure 2.2c   Forward and backward scheduling (slack added in backward scheduled items).

It is important to note that scheduling systems rely heavily on accurate time information and that actual lead times may change as workloads change because of the effect of queuing.

2.5.5 CAPACITY PLANNING

In the previous section, scheduling was discussed without reference to the availability of the resources needed to perform manufacturing tasks. The schedules shown in Fig. 2.2 are based upon the use of lead times which allow the correct start times to be calculated for each part. If the factory is not very busy, lead times will tend to be shorter, because the effect of queuing will be reduced, leading to stock waiting longer than necessary, and inventory will be higher than needed. On the other hand, high workloads increase lead times, so lead times could extend and delivery dates could be missed. If the high workload could be anticipated, decisions about rescheduling, overtime and subcontracting could be taken to remedy the situation, although at extra cost. If the workload could be taken into account when scheduling, it might be possible to avoid the need for such measures. This is the area of capacity planning.

The calculation of workload requires operation and set-up times to be known for each operation, and a capacity figure to be established for each work centre. The capacity of each work centre is usually established on the basis of the standard shift length less an allowance for regular maintenance.

The two alternative approaches are infinite capacity planning and finite capacity planning.

Infinite capacity planning

Infinite capacity planning (also known as *capacity requirements planning* (CRP)) looks at a schedule which has been produced by either forward or backward scheduling. For each period in the schedule (day, week or month) the load in working hours is totalled, using the operation and set-up time for each operation. The load can be compared against the capacity of the work centre, to identify any overload. Thus the capacity required during each period is identified.

Overloads can be dealt with by:

- increasing the capacity of the work centre (by overtime working or an extra shift)
- transferring some work to an alternative work centre or to a subcontractor

- as a last resort, concluding that the master production schedule cannot be achieved and must be replanned.

The only capacity problems that should arise at this level are because the sequence of work happens to overload a particular planning period. If the original MPS was planned with care, using RCCP, then the period which is overloaded at the capacity planning level should be part of a longer period on the master production schedule which is not overloaded. This means that there will usually be slack nearby in the capacity plan, which should allow some rescheduling at the capacity planning level without changes to the MPS.

Figure 2.3 shows a capacity plan which reveals an overload in the third week of April. As overloads were avoided at the MPS level, the overall month is not overloaded. The average load over the month is 90% of capacity. This means that besides being able to consider overtime and alternative resources, the production manager is able to consider using the available capacity in the preceding week to prevent the overload. However, this depends upon the work being made available to be worked on before it is due.

If this is not possible, it may be feasible to allow some of the spare capacity in the fourth or fifth week to be used. This means that work is behind schedule and will create disruption for the next work centre on its route. Infinite capacity planning does not provide any information to enable the extent of this disruption to be evaluated. Infinite capacity planning only reveals the extent to which specific work centres are overloaded.

Corke (1985) suggests that:

...the question that management really wants answered is: If we increase the work centre capacities by certain amounts, when will the various products be completed?

Finite capacity planning systems address this question.

Finite capacity planning

Finite capacity planning can use forward or backward scheduling. It works by actually booking time on the required work centres for each operation on each component or assembly. As the system looks forward or backwards in time it finds the next appropriate slot and books the amount of time required for the specific operation. There should never be an overload, because the system can only assign each operation to a time when the work centre is free.

Rough-cut capacity plan: work centre xyz

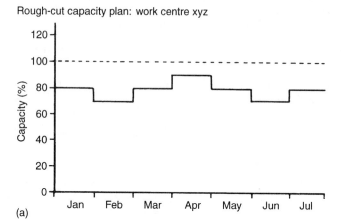

(a)

Capacity plan: work centre xyz

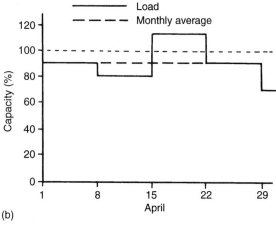

(b)

Figure 2.3 RCCP and feasible capacity plan.

Theoretically this means that there should be no queue time, because each work centre should be free at the appointed time when a scheduled workpiece arrives. In practice, however, extra time must be built in to allow for inaccurate estimates of operation and set-up times and for inevitable hold-ups. This can be done either by adding an allowance to all times or by using capacity figures slightly lower than the actual capacity available.

Rather than creating an excess workload in a given period, the system will keep on searching until an available time is found. This means that the lead time is a result of the capacity planning routine, rather than an input. If the lead times turn out to be too long, then either the finish times will be too late (forward scheduling) or the start times will be in the past or before material can be procured (backward scheduling). In either case the system can be re-run with an updated set of resources, such as to allow overtime, an extra shift or subcontracting. If after all the extra capacity is added the required dates cannot be achieved, the customer due dates in the master production schedule must be adjusted.

Because of the inherent complexity of the calculations, since every operation on every item must be allocated a time, and effectively a diary must be kept for each work centre (which must be searched when scheduling each operation) this kind of capacity planning system is notoriously slow.

In companies where there is a considerable degree of variation in product mix, the estimated machining times generally show some significant inaccuracy. These errors can build up to cause an appreciable over- or underload at the end of each day or week. Since these errors have knock-on effects to other operations, the errors can be compounded. Finite capacity planning is often regarded as being notoriously inaccurate.

2.5.6   FEEDBACK

The CAPM system must maintain information on the current state of resources and materials, so that plans can be updated if work does not progress as expected. Feedback is often achieved manually by checking the progress of work from time to time, but it can also be automated by the use of tracking systems, such as bar codes, which can be used to identify a product and which are read into a computer whenever the item passes through certain key stages of manufacture. Feedback is a key feature of closed-loop MRP systems, which are discussed in Chapter 4.

2.5.7   STOCK CONTROL AND PURCHASING

An essential function of CAPM systems is to ensure that materials are available as required for use in manufacturing or for sale to customers. This can be done either by maintaining stocks of each item to cover varying demand, or by purchasing and producing to meet the requirements of specific customer orders. The approach of maintaining stocks will be dealt with in Chapter 3. The alternative approach of issuing orders in line with demand, material requirements planning (MRP), is dealt with in Chapter 4.

SUMMARY

This chapter has described the goals of production management and some of the principles which are used. Production management systems may contain human activities or computer software to handle bill of materials, master production scheduling, rough-cut capacity planning, scheduling, capacity planning, stock control, purchasing and feedback. The following chapters will look at different approaches to the management of production, and the final chapter will consider the difficult task of developing a CAPM system to suit the needs of a particular business.

QUESTIONS FOR DISCUSSION

1. How is the performance of a company judged?
2. What measures may be used to evaluate the performance of a CAPM system?
3. Which of the three performance measures provides the greatest scope for improvement?
4. What are the functions of a CAPM system?
5. What elements may make up the lead time for the manufacture of a product?
6. What are the principal elements of a CAPM system? What is the role of each of them?
7. Define the terms 'load' and 'capacity'.
8. What is the difference between forward and backward scheduling?
9. What is the difference between infinite and finite capacity planning?
10. A capital goods manufacturer using a jobbing factory allows sixteen weeks for component manufacture and eight weeks for assembly, for any product in their wide range. What elements would you suggest would be required in the CAPM system?
11. A batch manufacturing company forward schedules all component manufacture and assembly work on the basis of allowing two operations per week. What data would be needed to determine whether lead times could be reduced? What alternative planning systems could be used?
12. Corke (1985) suggests that '...the question that management really wants answered is: If we increase the work centre capacities by certain amounts, when will the various products be completed?'. What types of manufacturing companies do you think Corke has in mind? Are there any manufacturing companies for whom other questions would be more important?

REFERENCES

ACME (1991) *Computer Aided Production Management*, Report of the research initiative funded by the ACME Directorate (Application of Computers to Manufacturing Engineering), Science and Engineering Research Council (SERC) (now Engineering and Physical Sciences Research Council), Swindon.

Corke, D.K. (1985) *A Guide to CAPM*, Institute of Production Engineers.

Fox, R.E. (1982) OPT – an answer for America (Part 2), *Inventories and Production*, 2(6).

Goldratt, E. and Cox, J. (1984) *The Goal*, Gower (revised 1986, 1989, 1993).

Waterlow, J.G. and Monniot, J.P. (1986) *A study of the state of the art in computer aided production management in UK industry*, ACME Directorate of SERC, Swindon.

Stock control 3

There are very few manufacturing businesses which manage to operate without stock, yet one of the key performance measures of manufacturing is to keep inventory levels (including stock) as low as possible while still being able to operate the business to satisfy customer demands. Many CAPM systems developed from stock control systems. This chapter aims to explain the way the production manager can establish a policy on stock control.

First, we shall look at the items that constitute stock. Three main kinds of stock can be identified: finished goods, work in progress and purchased items.

3.1 TYPES OF STOCK

Finished goods are held intentionally when a key market requirement is to be able to satisfy the customer with immediate delivery. Finished goods are held unintentionally when manufacturing items to a forecast which turns out to be incorrect.

In either case, the root cause of holding the stock is the extent of the company's inability to produce goods in the time the customer is prepared to wait. If the goods can always be produced in the time the customer is prepared to wait, there is no need to hold any stock of finished goods.

If demand is irregular and it is necessary to meet any demand from stock, then the company must hold in stock the number of items which could be demanded while more are made. This is expressed by the demand rate multiplied by the manufacturing lead time.

Since the stock is to cover demand which is irregular, any calculation is error-prone and can only give a rough indication based upon historical demand. It may also be required to hold some stock against possible hold-ups in manufacturing if the manufacturing lead time is also uncertain. For these reasons, safety stock is normally added to the calculated stock requirement.

Stated algebraically,

$$\text{Required FGS} = (D \times MLT) + \text{safety stock}$$

where:

FGS = Finished goods stock (pure number)
D = Demand rate, units per period (1/time)
MLT = Manufacturing lead time (time).

As the customer has a choice of supplier, there is a trade-off between the competitive benefits of short delivery times and the cost of capital invested in stock for an uncertain period.

Work in progress (WIP) is all the materials and parts which are either being worked on or are waiting for the next operation. If products are made and sold at any given rate, the amount of work in progress at any instant is proportional to the manufacturing lead time. This can be illustrated with a simple example.

Consider a product line which sells at the rate of one unit per day, and this unit is sold at the end of the day. On any particular day there must be one unit in progress which will be finished at the end of the day. If a unit can be made in a day, there is no need for any other unit to be in production until tomorrow. The level of work in progress is constant at one unit.

If the item takes two days to make, then at all times there must be one which is in its second day and which will be finished by the end of today, and one which is in its first day and which will be finished by the end of tomorrow. Thus the level of work in progress will always be two.

Stated algebraically,

$$WIP = LT \times D$$

where:

WIP = Work in progress (pure number)
LT = Lead time (time)
D = Demand rate, units per period (1/time).

If the lead time can be reduced by improved manufacturing techniques or organization, then besides being able to reduce the stocks of finished goods, the amount of inventory on the shop floor will be reduced.

Purchased items include both raw materials and purchased components. It may also include items which are bought simply to resell. (This is known as factoring.) Whereas finished goods are stocked in order to cover the time period between customer order and manufacturing, purchased items must be held to cover the time period between a need arising for items to be used in manufacture and the earliest they can be provided by a supplier. This time period includes the time taken to raise a purchase order and the time taken by the supplier to deliver the goods.

If demand for an item is constant, then an arrangement can be made with a supplier for regular deliveries, and no stock need be held (although most companies would still hold a small stock of important items, in case of hold-ups).

If demand is irregular, a company may take the decision to order materials only when an order for a product is received. If no stock is held, that means that the supplier lead time is added to the manufacturing lead time. If no finished goods stock is held, this means that a customer order can only be satisfied at the end of a long total lead time. If demand is irregular but customer lead time must be kept to a minimum, it may be decided to keep enough purchased items to cover the estimated level of demand during the time it takes for the supplier to supply the item. The number to be held in this case is the number which may be called for during the time between ordering from a supplier and receiving the delivery. This is given by the demand rate multiplied by the supply lead time, plus any safety stock needed to cover hold-ups or variations in demand.

Stated algebraically,

$$\text{Required PS} = (D \times SLT) + \text{safety stock}$$

where:

PS = Stock of purchased items (pure number)
D = Demand rate (1/time)
SLT = Supply lead time (time).

3.2 ORDERING

Stock is controlled by determining when to issue orders for the purchase or manufacture of items. To regulate finished goods stock, orders are released which specify what items are to be manufactured at any particular time. If the rate of production is higher than the rate of use (which here means sales), then the level of finished goods stock will increase. Conversely, if stocks are used up faster than they are supplied, stock levels will fall.

Low stock levels contribute to a high return on capital for the company. However, this is at the risk of losing orders if the stock level is so low that the lead time to customers is long, or if the company fails to supply on time as promised.

While the expressions above offer simple ways of calculating the amount of finished goods stock, work in progress and purchased items, it is not so simple in practice. In particular, the levels of finished goods stock and purchased items stock depend on the level determined for safety stock (sometimes called 'cover'). The level of safety stock can only be set by taking a view of the likely levels of orders which will be received in the future, and weighing the cost of the stock-holding against the cost of failing to meet some orders. Thus a stock-holding policy must be determined.

3.2.1 STOCK-HOLDING POLICY

In companies which operate at high volumes of production and with low variety, stock control is simplified for two reasons. Firstly, the high volume should mean that experience can be built up on the amount of variation in demand, which may be both random and cyclic (seasonal). Secondly, the low variety means that there are fewer items to hold in stock, so a policy of holding a certain amount of every item may be feasible, since the value of the stock and the number of items to store and record is comparatively low.

In companies which operate at the other extreme, of high variety and low volume, stock control is much more difficult. The irregular arrival of low numbers of customer orders, which may be substantially different, means that the level of demand for any particular item varies enormously, often from zero demand for a long period. The high number of items dealt with in such companies means that the cost of holding a stock of every item would be very high, simply because there are so many items, which in addition may each be expensive. These problems may be alleviated in part by attempting to use as many common parts as possible in the design of products, so that the required volume of each may become high enough to attempt to forecast aggregate demand. At the same time, the number of items for which stock-holding is required is reduced, so the stock problem is simplified.

3.2.2 LOAD SMOOTHING

Where demand fluctuates widely, the levels of stock which must be held tend to be higher, in order to try to satisfy the peaks of demand. Clearly, between the peaks the stock will stand idle, simply tying up the company's working capital. One answer is to try to stagger the demands so that some of the requirements of the peak can be dealt with in the trough. This is known as 'load smoothing', since it aims to provide a steady workload for the factory. The achievement of a smooth load simplifies stock control and also makes the operation of the factory more economical, since the need to use expensive extra capacity, such as overtime working or subcontracting, is avoided.

Load smoothing is the major purpose of master production scheduling and capacity planning. The need for a smooth load is a strong argument against combining orders into large batches, as production controllers are often tempted to do. A large number of smaller batches are much easier to schedule than a small number of large batches. A smooth pattern of demand makes stock-holding easier, since it removes the sudden peaks of demand which cause stock-outs and the troughs during which excess stocks are held.

3.2.3   THE ECONOMIC ORDERING QUANTITY

The cost of a particular stock-holding policy can be measured by assessing:

- the value of being able to meet any order (or the cost to the business of potentially losing an order)
- the cost of ordering items (raising paperwork, delivery costs etc.)
- the cost of holding the items in stock (cost of money which cannot be invested elsewhere, cost of storage space and stores staff).

If a demand rate can be assumed to be constant (which it usually can not) it can also be assumed that no orders will be lost through lack of stock, since the stock-holding policy can be formulated to ensure that the known demand can be met. In this scenario, therefore, the cost of lost orders does not arise. In this rare case, the cost of the stock-holding policy depends on the cost of ordering, which is lower if more items are held and less ordering is required, and the cost of holding the items in stock, which is lower if more orders are raised and less stock is held.

As an example, consider the stock-holding policy for an item with a value (V) of £10. We must assume that the demand for the item (D) is constant at a certain level, let us say 720 per year, and that the demand does not vary during the year.

We must also assume that the cost of raising an order (Co) can be calculated, and that the ordering cost and the price of the item do not vary with the quantity ordered (that is, no discount for order volume). Let us say that the cost of making an order and having the items delivered is £10. In practice, various means of apportioning cost to orders are used, which provide widely varying figures.

Box 3.1   Terms used in stock calculations.

Co Cost of making an order (£)
Cs Cost of investment in stock (proportion of stock value) (%)
To Total cost of making all the orders for the year (£)
Ts Total cost of investment over the year (stock-holding cost) (£)
V Value of the item (£)
D Annual demand for the item (pure number)
Q Quantity ordered in each order (pure number)
N Number of orders (pure number)
I Average stock level (inventory) (pure number)

We must also assume a cost of holding the items in stock (Cs), which for the sake of argument we could assume to be the lost benefit from not investing the money value of the goods in a way which would provide a return. (There is also the cost of the storage space, but let us assume that this is negligible.) Hopefully, the most profitable investment opportunity would be to expand the business itself, but as that opportunity is not always available the best alternative may be to put the money on deposit at the bank. (There is also the possibility of buying shares in a better manufacturing company, but we will discount this by assuming that we hope to make a success of our own company. If our company is traded on the stock market, this may not be true.) Interest rates from both banks and stock markets are constantly changing, but for the sake of this example let us assume that the interest lost would be at a rate of 10%. That means that the cost of holding the items in stock for a year is always 10% of the value of the amount of stock held. As the amount of stock varies while it is being used and replenished, the average value of stock held will be used.

The total cost of all the items used in the year (£7200) will have to be spent under any stock-holding policy, and can therefore be ignored. The stock-holding policy aims to minimize the total cost of ordering and investing in stock.

If the 720 items are purchased in one single order, the stock level during the year will be as shown by line a in Fig. 3.1, with an initial stock-holding of 720 being used at a constant rate until at the end of the year none are left. The average inventory (*I*) or stock level is 360. The total cost of holding the stock (*Ts*) is 360 × £10 × 10% = £360.

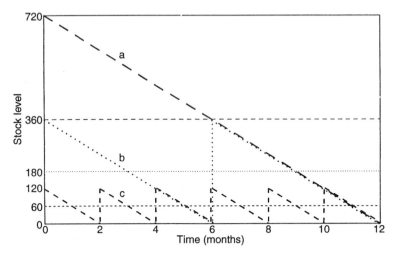

Figure 3.1 Inventory levels in three stock-holding policies.

This can be expressed algebraically thus:

$$Ts = IVCs$$

or since *I=Q/2*,

$$Ts = \frac{Cs\,V\,Q}{2}.$$

The cost of making the order is £10. If there were more than one order, the total cost of ordering would be given by *To=CoN*, where *N* is the number of orders. Since *N=D/Q*, this can be rewritten thus:

$$To = \frac{CoD}{Q}.$$

By adding the stock-holding cost *Ts* to the ordering cost *To*, it can be seen that the total cost of this policy over the year is £370.

Alternatively, if the 720 items are purchased in two orders of 360, the stock level during the year will be as shown by line b in Fig. 3.1. The average inventory *I* is 180. The total cost of holding the stock *Ts* is 180 × £10 × 10% = £180 and the total cost of making the two orders *To* is £20. The total cost of this policy over the year is £200.

Table 3.1 summarizes the costs of various stock-holding policies ($V=£10, D=720, Co=£10, Cs=10\%= 0.1$).

Table 3.1 Costs of various stockholding policies for a year's supply of 720 items worth £10 each. (a), (b) and (c) correspond to lines a, b and c in Fig. 3.1

Order quantity (Q)	Number of orders (N)	Average stock level (I)	Total cost of ordering (To) (£)	Total cost of stock holding (Ts) (£)	Total cost of policy (T) (£)
720 (a)	1	360	10	360	370
360 (b)	2	180	20	180	200
240	3	120	30	120	150
180	4	90	40	90	130
120 (c)	6	60	60	60	120
60	12	30	120	30	150

As the number of orders increases, the total cost of ordering increases, while at the same time the amount of inventory held decreases and therefore the stock-holding cost also decreases. This is shown graphically in Fig. 3.2.

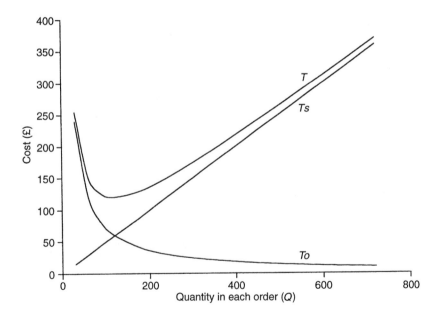

Figure 3.2 Costs of stock-holding and ordering.

The lowest-cost policy is found at the point where the curve of total cost is at a minimum, that is, where its gradient is zero. The total cost is given by:

$$T = \frac{C_sVQ}{2} + \frac{C_oD}{Q} \, .$$

Differentiating with respect to Q gives the gradient of the total cost curve:

$$\frac{dT}{dQ} = \frac{C_sV}{2} - \frac{C_oD}{Q^2} \, .$$

The lowest point is where $dT/dQ = 0$, i.e. where

$$\frac{C_sV}{2} - \frac{C_oD}{Q^2} = 0 \, .$$

We can rearrange this expression to give the order quantity (Q) which corresponds to the lowest cost thus:

$$\frac{C_sV}{2} = \frac{C_oD}{Q^2}$$

$$Q^2 = \frac{2C_oD}{C_sV}$$

$$Q = \sqrt{\frac{2C_oD}{C_sV}} \, .$$

This value of Q is often referred to as the *economic ordering quantity* or *EOQ*. It suggests that the most economical stock-holding policy for the item in the example above is to purchase the items in lots of 120, thus ordering the parts six times per year or once every two months. The stock level for this policy is shown by line c in Fig. 3.1.

In this example, a year's demand was considered, together with a year's interest charges. However, it may be noted that since the demand D and cost of stock-holding C_s are both expressed in units per time period, the time period is cancelled out and may be ignored. Thus the expression for EOQ is equally true for any period of time as long as the demand is known for the period and the cost of stock-holding is known for the period. If the stock-holding cost is given for a year, as usual, then a demand rate for a year must be calculated. If a production program demands 500 items over a six-month period, then an equivalent annual demand of 1000 items should be used for the value of D in the equation.

There are some drawbacks to the use of the economic order quantity. While the arithmetic is sound, let us reconsider the assumptions that have to be made:

- that the demand rate is constant
- that no customer's order will be lost through lack of stock
- that the cost of raising an order (Co) can be calculated
- that the ordering cost and the price of the item do not vary with the quantity ordered
- that the cost of holding the items in stock (Cs) can be calculated, and will remain constant
- that items are paid for when they are bought or made
- that items can be bought or made instantly
- that items can be bought in the quantities required
- that items can be considered individually (ignoring the fact that they may only be used together with certain other items, which appear to require a different ordering regime)
- that stock can be allowed to run down to zero with no safety stock.

Since demand is assumed to be constant, the EOQ balances the cost of ordering against the cost of holding stock. It takes no account of the possibility that demand for an item may cease completely, such as in a design change, in which case the company will be left with a rather uneconomic stock of spare parts. In order to make the company as efficient as possible in stock-holding, it should be attempted to reduce the levels of stock and the factors which lead to stocks being held. The most troublesome assumption made in the EOQ calculation is that the cost of ordering cannot be changed. If in the example above the ordering cost was reduced from £10 to £5, the EOQ would fall from 120 to 85 and the amount of money tied up in stock would be reduced by almost 30% from £600 to £425. Improved relationships with suppliers, long-term supply contracts, electronic data interchange and reduced paperwork are some possible means of reducing the ordering cost.

3.2.4 ECONOMIC BATCH QUANTITY (EBQ)

A variation of the economic ordering quantity is sometimes used to determine the number of items to be produced in the factory in one order, that is, the batch size. The economic batch quantity corresponds to the economic order quantity, the main difference being that the cost of ordering is replaced by a term which includes both the cost of making the order and the cost of the set-up times for the machines concerned. The set-up time cost is assumed to be the greater and in EBQ calculations the cost of raising the order is often disregarded. Whereas EOQ balances inventory against the cost of raising orders, EBQ balances inventory against the cost of resetting machines.

Besides the assumptions of EOQ described above, EBQ requires us to assume:

- that the cost of setting up machines can be determined (besides the labour cost of the time spent, there is a cost associated with not using the time to make products which would generate cash through sales, while if the machine has some free time available, it may as well be used for a set-up, and the set-up may have no cost)
- that the cost of setting up machines cannot be reduced
- that the value of each item can be determined (an item can be said to be more valuable after it has been worked on, but the amount invested in an unsold item is related to the cost of not having used the factory to make something else).

Most importantly, the EBQ calculation invites the production manager to treat the set-up time and cost as fixed, and to use them to determine a batch size policy. An alternative view is to find ways of producing items either continuously, with no set-up costs, or in very small batches with such low set-up costs that production can operate with a very low level of work in progress inventory. This view places the emphasis on improving the factory's skills, particularly in finding good engineering practices to reduce set-up times and to increase the flexibility of machines and equipment.

Like EOQ, EBQ continues to be used despite requiring assumptions to be made which are seldom true.

3.3 MORE REAL-LIFE ASPECTS

Stock-holding policy must take account of the following factors: variation of demand, lead time needed to secure more stocks and variation of lead time needed to secure new stocks.

In most cases, stock cannot be provided instantly, but only after a supply lead time for purchased items or a manufacturing lead time for items made in-house. Stock is used to allow for this lead time, and safety stock is added to cover its variation. To prevent a shortage, enough safety stock must be held to cover the length of time needed to procure new stock. If it takes a month from the time an order is issued to a supplier to the goods being delivered and ready to use, then new stock must be ordered while there is enough remaining to cover the requirements which are expected in the month. If demand is constant, that simply means making the order a month early. However, as demand tends to vary, it may be that shortages and excesses will occur if the actual demand is not taken into consideration. This can be achieved in two ways.

Stock replenishment sets a stock level which is equal to the EOQ plus the amount of safety stock required, and at the normal reordering interval (Q/D years) the stock level is checked and an order is issued to bring the stock level up to the desired level, thus replenishing that which has been used. This has the advantage that the supplier will know when to expect an order. Over a length of time, the average amount ordered should approximate to the EOQ.

Reorder point calls for the stock level to be monitored, and an order is raised for the EOQ as soon as the stock level falls to a certain point known as the *reorder point* or *ROP*. The required stock level is therefore maintained by orders of regular quantities but at irregular intervals. At the reorder point, enough stock should remain to cover the demand expected during the supply lead time, which may itself be variable.

A common form of reorder point stock control is the two-bin system. This system approximates the economic order quantity by setting the safety stock equal to the order quantity. This quantity of items is held in a bin or container of suitable size. One bin is made available for use in the factory, while an identical bin is kept in the stores. When the bin in use is empty, it is exchanged for the full one, and at this point an order is raised for the same quantity of items from the supplier, or from the factory itself if the item is manufactured in-house. Thus the safety stock is equal to the order quantity, and because the system only works if the items can be purchased in the time taken to use the bin quantity, the quantities tend to be large. This means that the system is not suitable for high value items. The two-bin system is similar to the *kanban* system described in Chapter 5, which may be thought of as a many-bin system.

3.4 STOCK-HOLDING POLICIES FOR DIFFERENT KINDS OF ITEM

So far, this chapter has made the assumption that stocks of items can be maintained. If demand is very low or very variable, and especially if items have high value, the cost of stock-holding against unknown demand may make the holding of stocks very undesirable. Of course, it is possible to purchase items only when they are needed to satisfy customer orders, or when there is a good forecast that they may be needed. Each company must decide which items should be stocked to ensure their availability and which should be ordered only as required. A common classification is to divide purchased items into A, B and C items. This analysis can equally well be applied where the items are manufactured in-house.

This classification depends upon the value of the item and upon the frequency with which it is required. If an item is of high value, it is more expensive to hold stocks. If an item is required very often, it may not be practical to make an order every time one is required. In order to take both these factors into account, the *usage value* or *value of use* of the item may be calculated. This is simply the value of the item multiplied by the amount used in a given time period, such as one year. A high value of use means that the item is significant and must be carefully managed. This may mean that the item should be purchased as required, or in accordance with the most careful forecasting. A low usage value means that, with respect to other items, this item is relatively insignificant and may be ordered according to an EOQ based upon the historical use of the item. The absolute values concerned will vary between companies. This analysis provides a way to determine which items deserve more attention in stock-holding and purchasing, and can only be used to compare items within one company.

A-items have a high usage value, and usually a high value. The major subassemblies of a product are likely to fall into this category, and their value will be a high proportion of the value of the product. A-items are often purchased only when required by a specific customer order. However, if the time available to purchase the item is often less than the supply or manufacturing lead time, the item must be purchased according to a forecast, so that an item will be available to satisfy the order when it is received, without delaying the delivery of the finished product to the customer. If the forecast turns out to be wrong, then a large investment has been used for no benefit, and the company's inventory value is increased. Improvement activities can attempt to deal with this in two ways. By improving communication channels to the customer, the company can more correctly assess the likelihood of orders being received, and can improve its service to the customer by being in a better position to satisfy the order. By improving communications with suppliers, the company can help the suppliers to know what orders to expect, which will allow them to obtain necessary items. This may also help the suppliers to reduce their lead times, thus reducing the need to work to a forecast.

B-items include high-value items which are required less often and low-value items which are required very often. B-items range from those items which border on the A-class and are almost too expensive ever to have in stock, to those which are almost C-items and are cheap to hold. B-items are the range of items in the middle, where for any particular item a stock-holding policy must be determined. While it represents probably the largest group of items, the B-class is difficult to define. Work at IBM (Shah *et al.*, 1990) has shown how the B-items can be classified into groups, each with a different number of orders per year (N) calculated according to the economic order quantity. In this system, an MRP system (Chapter 4) was used to determine the amount in each order, so that ordering would only be for expected requirements, rather than for the actual economic order quantity. The B-items with the highest value would be ordered most frequently.

C-items have a low usage value, and usually a low item value. Their EOQ would be expected to be high. In some cases the EOQ can be so high that the items could be made obsolete by design changes or material shelf-life, which might necessitate more frequent ordering. The two-bin system is often suggested as a solution for C-items. While less attention may be focused upon C-items because of their low value, it is important to realize that a shortage of a C-item is just as serious a hold-up in manufacturing as a shortage of any other item. This is often dealt with by allowing generous safety stocks, since the cost relative to other items is low. An alternative may be to work with suppliers to reduce the cost of ordering through long-term call off arrangements. This is dealt with in Chapter 5. Companies who manufacture high-value products which contain C-items, possibly in quite high numbers, may wish to ensure that they do not actually make any of the C-items, so that their resources are not wasted in the production and control of items of low added-value.

3.5 USE OF COMPUTERS

Most modern CAPM packages provide applications which can help with stock control. The reorder point approach is commonly used to deal with the management of stock levels and economic order quantities, issuing orders as the reorder point is reached, although this requires the stock level to be recorded each time stock is used or received. Computerized reorder point is thus highly dependent on good data accuracy. ROP can be operated manually in situations where the number of different items is small, or where the company simply deals with a supplier company who effectively takes over the stock control problem. The replenishment system, where orders are raised at intervals according to past use, is often operated manually by, for example, a weekly issue of orders for whatever items are required. This can also be computerized, although since the computer has nothing else to do but raise orders it can equally well issue them at the reorder point or at the replenishment interval.

For the effective management of large numbers of items for which demand is low, computerized systems are able to keep track of the current stock levels, supply and manufacturing lead times and the bill of materials for each product, processing thousands of items in a relatively short time. Many of these systems combine reorder point ordering for stock items with MRP ordering for items purchased to meet specific orders. MRP is dealt with in Chapter 4.

SUMMARY

This chapter has described the relationship between the variables involved in stock control. While stock control is a critical part of production management, companies often rely on rules of thumb, such as the economic ordering quantity, which have very limited applicability since the assumptions which they require are rarely true in any real-life situation. The job of the production manager, therefore, should be to challenge the assumptions made in the business to remove those factors such as set-up times or ordering costs which lead the company to set uneconomical policies. The aim should be to be able to satisfy a customer order for one item in the time the customer is prepared to wait and without investing in more stock than is required to satisfy the present order.

QUESTIONS FOR DISCUSSION

1. What kinds of stock may be held by a manufacturing company?
2. Why is each kind of stock held?
3. What costs are balanced by the economic ordering quantity?
4. What is the difference between stock replenishment and reorder point stock-holding policies?
5. What are A-items, B-items and C-items?
6. In what kinds of industries are companies able to hold very low stocks of finished goods? What are the key characteristics of these industries? Why do some companies hold higher levels of finished goods stocks?
7. A simple product, a garden windmill, sells at a rate of about ten per week, for a price of £80. It is made up of two main components, which are purchased at £12 and £18 each. The cost of raising a purchase order is estimated at £24 and the current interest rate is 8%. The value of the work and small components required to complete the assembly is approximately £30, and the cost of ordering a product to be finished is considered negligible. What stock-holding policies would you suggest for each of the components and for the finished product? Calculate the inventory value and stock turn ratio for your policy. What assumptions have you made in determining your answer?
8. If an improved stock-holding policy reduces the cost of ordering, where does the saving show up in the company's accounts?

REFERENCE

Shah, S., Burcher, P. and Relph, G. (1990) Extending the Pareto principle to MRP controlled parts and regaining MRP control, *BPICS Control*, April/May, 39–45.

MRP, closed-loop MRP and MRPII

4

4.1 INTRODUCTION

MRP has become a general term used to describe material requirements planning (MRP), closed-loop MRP and manufacturing resource planning (MRPII). This chapter will attempt to explain how each works and the differences between them.

One of the principal authorities on the subject was the late Oliver Wight (1981). No description of the various MRP approaches can be complete without reference to his work.

MRP is the basis of the vast majority of CAPM systems, and the term 'MRP' is sometimes used loosely to refer to the CAPM systems of a particular company, which may include various other functions besides MRP itself.

The advent of cheap computing power in the 1950s and 1960s led to a rapid growth in the number of companies using MRP. After accounting and payroll systems, MRP was one of the earliest computer applications to become widespread. MRP offered companies the ability to order goods according to actual requirements, rather than simply replenishing standard stock levels. Initially, MRP was used mainly for purchased items, but the extension of its use to include in-house manufactured items followed, especially in those factories where there was irregular demand and a wide range of items, such as in the aerospace industry.

According to Browne *et al.* (1988), MRP and MRPII have been the most widely implemented large scale production management systems since the 1970s, with several thousand implementations worldwide.

4.2 SIMPLE MATERIAL REQUIREMENTS PLANNING (MRP)

4.2.1 TIME-PHASING

Callerman and Heyl (1986) state:

> MRP is a time phased order release system that under ideal circumstances schedules the order releases for needed demand inventory items so that the items arrive just as they are required

Simply stated, MRP systems calculate the date upon which an item may be ordered either from the factory or from a supplier, by reference to the date on which the item is needed and the time that it is expected to take before the item arrives. This is a very simple operation which is easy to perform manually for a limited number of items. However, in order to perform this simple function, it is also necessary to identify the exact items required and the dates on which they must arrive.

MRP operates by backward scheduling each item from its requirements date to launch production and purchase orders accordingly so that they will be completed on time. The main feature of the MRP family is that MRP uses a lead time for each item. The lead time for each item may be held in a file often known as the item master file (Box 4.1).

Box 4.1   The files used by MRP systems.

BILL OF MATERIALS (BoM)

The BoM file is a list of all the purchased parts and raw materials required to manufacture a finished product. It can be broken down into subassemblies which in turn require other purchased parts and raw materials.

MASTER PRODUCTION SCHEDULE (MPS)

This file, as its name implies, stores the company's plan of what to manufacture and when. It is derived from the orders received from customers and forecast orders. The accuracy of this file together with its efficient update as required is fundamental to the effectiveness of the MRP system.

INVENTORY STATUS FILE

The inventory status file keeps a record of the amount of stock and usually of stock transactions. As in the case of the MPS, the accuracy of this file is critical. This is because when the BoM is exploded and requirements for the next time period are calculated the amount in stock is subtracted from the gross amount required, giving the net amount required to be produced. Clearly, if the amount in the inventory status file is incorrect then an incorrect amount will be manufactured. This can result in excess stock and shortages.

PARTS MASTER FILE/ITEM MASTER FILE

This file holds the information relevant to ordering each part. The file normally stores the standard batch sizes, lead time and any stock categorization which the part may have, and sometimes the routing information. A separate file may be used for the details of purchased parts. In some implementations the routings may be held separately in the routing or process file.

ROUTING FILE/PROCESS FILE

For each manufactured item, this file gives the sequence of operations and work centres used. If finite capacity planning was in operation then predicted or standard times for each operation would be stored here.

WORK CENTRE FILE

This file stores details about each work centre, such as costs associated with it, standard set-up times and time available on it (in cases where different shifts or maintenance periods apply to different machines).

TOOL FILE

This file can be used in connection with the work centre file and the parts master file. It stores details about tools, jigs etc. which are required for particular operations. This allows the MRP to order tools required for a specific operation.

For example, if all the components for a batch of 50 widgets are required for assembly at the start of June, and all the parts have a lead time from order to delivery of one month, then the orders must be released at the start of May.

In practice it is very unlikely that all the parts will have the same lead time, but the particular time for each item is stored in the item master file. The MRP software will release each works or purchase order from the parts list of each product the appropriate time before the requirement date for the specific order.

4.2.2 BILL OF MATERIALS

Since ordering is usually for items required to make up a product to satisfy a customer's order, MRP systems almost invariably provide the means to store and manipulate the list of parts required for each product. This is generally known as the bill of materials (Box 4.1).

4.2.3 INVENTORY MANAGEMENT

Because items may be held in stock under various different policies, MRP systems almost always provide the ability to record details of current stock levels and of orders issued for which the goods are not yet in stock, and items which are in stock but which are already allocated to be used for a particular customer order, and which are not therefore available for use.

4.2.4 DEALING WITH ORDERS

The addition of these functions, where they are needed, makes for a set of repetitive clerical tasks which lends itself to being performed by a computer. It is usual for the MRP system also to print out the actual paperwork which is sent to suppliers or to the factory, or to transmit the information by electronic means, such as electronic mail.

Purchase orders may take a variety of forms according to the contractual agreement between the supplier and customer. It is fairly common practice for purchase orders to be printed on multi-part forms which carry a standard set of purchase conditions. Various leaves of the form are retained for use in tracking the progress of the order and for checking against invoices and delivery notes.

Works orders, known variously as batch cards, travellers, route cards, dockets etc., identify the item to be manufactured, quantity and due date, often together with details of the manufacturing operations required, expected or allowed time, quality requirements, quality certificates to be completed, drawing of the item etc.

Many companies have found that supplier development has allowed them to simplify purchasing procedures and to reduce paperwork.

4.2.5 OPERATION

The operation of an MRP system is shown in Fig. 4.1. This will be explained through the use of an example product, an electronic calculator.

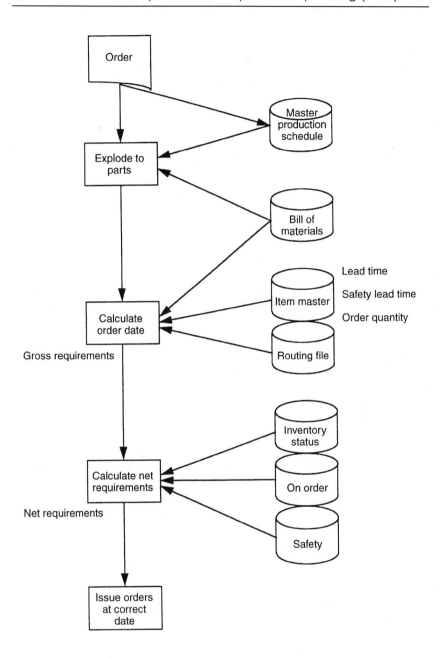

Figure 4.1 Schematic of MRP.

Customer orders, both firm and forecast, are listed in the master production schedule. In this example, the company only has two orders, both for the same product (Box 4.2). The parts list is held in a file called the bill of materials or BoM. (Bill of materials is the American term for parts list, which has passed into common use in MRP.) The BoM file is often printed in the indented form shown in Box 4.3.

Box 4.2   Master production schedule.

			Week		
Orders for calculator packed	23	24	25	26	27
A11			100		
B12					50

(Nothing else has been ordered!)

Box 4.3   Bill of materials (complete).

```
Calculator packed
      Calculator complete 1
            Lower case assembly 1
                  Lower case moulding 1
                  Battery cover 1
            Upper case assembly 1
                  Upper case moulding 1
                  Transparency 1
                  Adhesive badge 1
            Circuit board assembly 1
            Keypad 1
            Screws 3
      Carton 1
      Insert 1
      Instruction leaflet 1
```

This makes it easy to see how the product is constructed. For each item, the parts of which it is made up are listed beneath it, but indented one position further. The finished product is a calculator, packed and ready to be sold. Thus the finished product consists of the calculator complete, the carton, insert and instruction leaflet. The complete calculator is made up of the lower and upper case assemblies, the circuit board assembly, the keypad and three screws. The upper case assembly is made up of the upper case moulding, the transparent window and the adhesive badge. The lower case assembly is made up of the lower case moulding and the battery cover. (We will simplify the rest of the example by only dealing with some of these items.)

Existing and new orders are dealt with by the bill of materials processor within the MRP software. This uses the parts list for each product to produce a list of all items which must either be purchased or manufactured. This is known as the bill of materials explosion (Box 4.4). The list produced here is often enormous and contains details of the item and the customer order which needs it.

Box 4.4 Bill of materials explosion (some items not shown).

Item	Order	No. required
Calculator packed	A11	100
	B12	50
Calculator complete	A11	100
	B12	50
Carton	A11	100
	B12	50
Circuit board assembly	A11	100
	B12	50
Keypad	A11	100
	B12	50
Lower case assembly	A11	100
	B12	50
Screw	A11	300
	B12	150
Upper case assembly	A11	100
	B12	50

The item master file (Box 4.5) must now be referred to in order to calculate the date each item is required, using the due dates of the relevant customer order(s) and the lead time of each item above the item in the bill of materials. For the calculator to be packed in time for the order to be despatched at the start of week 25, the calculator and the packing materials must be available at the start of week 24, because the 'calculator packed' is allowed a lead time of one week. The keypad is used in the 'calculator complete' which has a lead time of one week, and the 'calculator complete' is used in the 'calculator packed', which also has a lead time of one week. Thus the keypads must be available two weeks before the date the customer order is due. For any part, the requirements date is found by subtracting the lead time of all the items in which it is used from the date the order is required.

Box 4.5   Item master file (some items not shown).

Item	Lead time	Make/buy
Calculator packed	1	M
Calculator complete	1	M
Carton	3	B
Circuit board assembly	4	B
Keypad	4	B
Lower case assembly	1	M
Screw	1	B
Upper case assembly	1	M

Box 4.6 Gross requirements (some items not shown).

Item	Week 22	23	24	25	26	27
Calculator packed				100		50
Calculator complete			100		50	
Carton			100		50	
Circuit board assembly		100		50		
Keypad		100		50		
Lower case assembly		100		50		
Screw		300		150		
Upper case assembly		100		50		

This provides the gross requirements. These are shown in Box 4.6. This describes the total number of each item which will be needed to satisfy all the orders in the master production schedule, according to the date on which each is required.

In some circumstances, where quality is unreliable, it is at this stage that extra items may be included by increasing the gross requirements to cover for rejects. If this is the case, it is to be hoped that efforts will be made to improve the quality, rather than allow uneconomic waste to persist. However, it is a feature of some MRP systems that a scrap allowance is shown in the item master file, thus providing an automatic way of increasing the gross requirements to cover up quality problems. It is a shortcoming of MRP that such features allow poor practice to continue.

The date at which the items must be ordered is not considered yet, as existing stock and orders which have already been placed must be dealt with.

The gross requirements are also shown in Box 4.7, which summarizes the MRP calculations. Once the gross requirements have been established, they are checked against the inventory status file or stock file to determine whether any of the requirements can be supplied from stock. If they can, these items must be identified as 'allocated' in the stock file to ensure that they are not seen as still being available by the next set of calculations. In Box 4.8 it will be seen that 100 screws are in stock (often described as 'on-hand'), but 40 of them have already been allocated to another product. This means that only 60 are available for use on this product, which is the number shown on the summary in Box 4.7.

Orders which have already been issued and which will arrive before the date of the gross requirements must also be considered. Even when pure MRP is used, with no intentional stocks and with no items being managed on a reorder point basis, orders tend to be made for convenient numbers and for economic order quantities (as described in Chapter 3). This means that stocks of some extra items may develop, which may reduce the net amount which must be ordered. In the example, 100 circuit board assemblies have been ordered and are expected in week 24.

The available stock figure and the on-order figure are subtracted from the gross requirement to provide the net requirement. The procedure for calculating the net requirements is called 'Gross-To-Net' or 'Netting-Off'. If there is more stock available than the gross requirements, such as for cartons in week 24, the excess stock will show in the 'available' row after the date of the gross requirements.

Box 4.7 Summary of MRP information (certain items only).

Item		Week								
		19	20	21	22	23	24	25	26	27
Calculator packed	Available									
	On-order									
LT=1	Gross							100		50
	Net							100		50
	To-order						100		50	
Calculator complete	Available									
	On-order									
LT=1	Gross							100		50
	Net							100		50
	To-order						100		50	
Carton	Available	200	200	200	200	200	200	100	100	50
	On-order									
LT=3	Gross							100		50
	Net									
	To-order									
Circuit board assembly	Available	10	10	10	10	10		100	50	50
	On-order						100			
LT=4	Gross					100		50		
	Net					90				
	To-order	90								
Keypad	Available	30	30	30	30	30			150	150
	On-order							200		
LT=4	Gross					100		50		
	Net					70				
	To-order	70								
Lower case assembly	Available	35	35	35	35	35				
	On-order									
LT=1	Gross					100		50		
	Net					65		50		
	To-order				65		50			
Screw	Available	60	60	60	60	60				500
	On-order								500	
LT=1	Gross					300		150		
	Net					240		150		
	To-order				240		150			
Upper case assembly	Available	20	20	20	20	20				
	On-order									
LT=1	Gross					100		50		
	Net					80		50		
	To-order				80		50			

Box 4.8   Inventory status file (certain items only).

Item	On-hand	Allocated	Available
Calculator packed	0	0	0
Calculator complete	0	0	0
Carton	200	0	200
Circuit board assembly	10	0	10
Keypad	30	0	30
Lower case assembly	35	0	35
Screw	100	40	60
Upper case assembly	20	0	20

The net requirements can now be converted into the requirements which must be ordered by moving backwards in time by the lead time for the item concerned. Thus the 70 keypads which are required in week 23 will be ordered in week 19.

It can be seen that 500 screws have already been ordered for week 26. Perhaps this is part of a regular ordering arrangement, but as the lead time for screws is only one week, it should be possible to bring forward the order to cover both the 240 and the 150 that will be required in weeks 23 and 25. If this is not done, it looks as though the 500 screws will go forward as excess stock. This can be checked by the use of a 'pegging report'. This is generated by the MRP system by tracing back through the MRP calculations to identify the requirement which led to the order. In this case it may show that the screws will be required for an order due in week 30, beyond the range of our MPS extract.

The ordering requirements for each item are shown in Box 4.9, which can be translated into an ordering schedule as shown in Box 4.10. This shows the orders in date order.

The orders will only be released when the correct time arrives. However, it is possible to alter manually the dates and quantities of the orders in advance, as may be required for the screws. This feature can be used to override the normal lead times or calculated net requirements when the production manager is aware of an unusual situation. This provides the production manager with the means to take control back from the computer, although it is possible to cause a large amount of disruption by altering the data if the full effects of changes are not known.

Box 4.9   Ordering requirements (certain items only).

Item	Week								
	19	20	21	22	23	24	25	26	27
Calculator packed						100		50	
Calculator complete					100		50		
Carton									
Circuit board assembly	90								
Keypad	70								
Lower case assembly				65		50			
Screw				240		150			
Upper case assembly				80		50			

Box 4.10   Ordering schedule (certain items only).

Week	Item	No. required	Week due	Make/buy
19	Circuit board assembly	90	23	M
	Keypad	70	23	B
20				
21				
22	Lower case assembly	65	23	M
	Screw	240	23	B
	Upper case assembly	80	23	M
23	Calculator complete	100	24	M
24	Calculator packed	100	25	M
	Lower case assembly	50	25	M
	Screw	150	25	B
	Upper case assembly	50	25	M
25	Calculator complete	50	26	M
26	Calculator packed	50	27	M

4.2.6   NET CHANGE AND REGENERATIVE MRP

All the early MRP systems operated on the basis which is now known as 'Regenerative MRP'. In this system, inventory, BoM and MPS files are updated during a production period, and the orders which are worked on and released during the period are those which were calculated at the start of the period. At the end of the period, the MRP program will be run to determine all the requirements based on the latest state of the MPS, BoM and inventory. Orders that have not yet been released are overwritten by new ones if the data has caused them to be incorrect, such as if a new customer order has appeared on the MPS, which will require more components to be bought. Orders may be released during the period between MRP runs, but they will only be as up-to-date as the state of the information when the MRP was run. Regenerative MRP is fairly inefficient in terms of data processing, since it often repeats calculations which have already been made. Early implementations could often only run MRP fairly infrequently, since the program would often require several days to run and would prevent other applications from using the computer during that period.

A more modern development of MRP is the 'net change' approach. This is a more sophisticated approach which leaves existing planned orders in place, but updates them if required by a new order or a change in requirements. Some net change systems can run in real time, calculating the effects of any change as soon as it is keyed in. Others achieve good results by performing a daily update, calculating the effects of all the changes made that day. The calculation requires much less computer power since only the items which have changed need to be dealt with.

4.3 DRAWBACKS OF MRP

Unfortunately, the straightforward MRP procedure relies naively upon the correctness of the lead time stored for each item. For purchased items, where suppliers quote delivery dates and strive to achieve them, particularly where the company has a good relationship with the supplier and the supplier is familiar with the item through frequent orders, it is possible to have a high degree of confidence in the lead times and the system can operate well. On the other hand, in-house production often has to be adjusted and rescheduled as the changing demands of customers alter the mix and level of workload in the factory. This can mean that the circumstances in which an item is produced can vary greatly from previous experience, with the result that lead times recorded in the file may not be a good guide to the likely lead time of an order issued today or next month. In particular, the level of load on the factory at any instant has a considerable effect upon lead times.

When the factory is busy, more work in progress means that parts generally have to queue before operations. This results in an increase in the overall lead times of all items. When the load is light, lead times may be shorter than expected, with the result that items are completed earlier than expected, when they may have to wait for purchased items to arrive before the finished product can be completed, thus meaning that inventory is higher than it need be for the amount of throughput.

The basic MRP system has no means of taking these variations into account, beyond placing the best estimates of lead time in the item master file.

Capacity planning, described in Chapter 3, can help by giving warning of overloads which can be used to provide extra capacity such as overtime working to prevent lead times being seriously extended by the changing workload. Capacity planning is available in many CAPM packages which work on the basis of MRP, but capacity planning is not part of the function of MRP itself.

The lack of capacity planning restricts basic MRP systems to only the simplest production systems. For example, where very similar parts are made and the variety does not affect volume of output, MRP could be used for routing, and simple counts of volume could be used for scheduling the correct amounts of work. Where lead times can be relied upon, such as in some purchasing arrangements, MRP can operate very satisfactorily. If demand is regular and the product mix is fairly stable, MRP may be no better than a stock replenishment system.

Data accuracy is very important to the success of an MRP system. Browne *et al.* (1988) believe that:

> Perhaps the greatest requirement of all for successful MRP installation and operation is discipline. This includes the discipline to maintain accurate stock records, the discipline to report accurately and in good time the completion of jobs and orders and to report to the system every event that MRP should be aware of. ...Many successful MRP installations have padlocks on the doors to the stockrooms.

If everyone has free access to stockrooms, there is always the possibility that items will be taken out without their issue being properly recorded, whether by mistake, by lack of familiarity with the records or to quietly replace a damaged item.

Browne *et al.* go on to say that one of the main causes of MRP failure is:

> Inaccurate data, particularly BoM data and inventory data.

Incorrect BoM data causes the wrong gross requirements to be calculated. BoM data becomes incorrect when design changes or component substitutions are implemented without being properly recorded.

Similarly, incorrect inventory data causes incorrect net requirements to be calculated. Incorrect inventory data can come from items being miscounted, scrap not being accounted for, or from items simply being lost.

At the time of assembly this can mean that parts which are not required have been manufactured or that parts which are required have not been manufactured.

If the lead time information is incorrect then the works orders are issued at the wrong time. This results in parts being completed late or in excess waiting time, which is wasteful.

When MRP is used to control both purchasing and in-house manufacturing, the company begins to depend upon the correctness of the orders issued by the MRP system, since the calculations involved are too complicated to perform by hand. Where the machine is being trusted to such a great extent, it is vital to ensure that the data on which it operates is as correct as can be managed.

4.4 CLOSED-LOOP MRP SYSTEMS

Oliver Wight (1981) sums up the meaning of 'closed loop':

> The term 'closed loop' really has two meanings. It means that the missing elements in the system like capacity planning, shop scheduling, and vendor scheduling were filled in. It also means... that there must be feedback from the vendors, from the factory, from the planners, etc. whenever there is any problem in executing the plan.

4.4.1 CAPACITY PLANNING

Standard MRP systems take the finished goods requirements and calculate the net material requirement. Closed-loop systems perform the same MRP function but then go on to produce capacity plans. The capacity plan attempts to show the load required on each work centre during each production period, whether it be a month, week or day. The capacity plan is calculated by using the routing file to show the workload which will be generated at each work centre by the net requirements. This shows the totals of the expected set-up and run times for all the items to be produced during the period.

4.4.2 CLOSING THE LOOP

Closing the loop in an MRP system involves providing the system with feedback about progress. Once an acceptable plan has been established, the system monitors progress against the plan. This is done in terms of the expected times of the work done. Thus a job which is expected to occupy a work centre for one hour is shown as a load of one hour in the capacity plan, and a value of one hour's worth of work is considered to be achieved when the job is finished, irrespective of how long the operation really took. In this way it is possible to measure progress against the plan. Such measures of time for performance assessment are known as 'standard hours'.

This feedback necessitates the recording of work as it progresses. In some factories, this is done by each operator booking the start time on beginning a job, and then booking off at the end, the computer keeping track of the time taken. Some of these reporting systems are quite sophisticated, requiring operators to book separately for each type of activity, such as set-up, run time, delays of various kinds and visiting the toilet. This degree of monitoring gives the impression that the computer can take account of all eventualities, as long as it has the right information, thus taking away the need for the operator, and often the supervisor, to worry about the progress of work. Many people resent being monitored in this way, and the accuracy of reporting is not always high.

A drawback of this kind of measurement is that the performance measured often does not take into account whether the work has been performed in the correct sequence. The MRP logic assumes that jobs will be worked upon when the orders are released, but if a particular item is delayed, and another is worked upon, the standard hours measure may show that performance is as planned, thus failing to identify the problem. This can have the result that the factory achieves its standard hours targets, but may not have produced all the items needed to assemble a product.

Closed-loop systems also update the system lead times. The latest recorded lead time may be used, or the program may calculate a moving average which is adjusted by each new figure. In plants that are always very busy, this allows the lead time to get longer and longer. The level of work in progress inventory (WIP) therefore increases, which in turn leads to increased queues. In order to ship the most urgent orders they must be expedited, thus delaying other work. Circumstances sometimes develop in which the factory can never produce anything until it is late and the expeditors chase it, because if it is not late it will be ignored while late work is dealt with. If the situation is bad, it means that non-late parts are ignored until they become sufficiently late to be top priority. This may be seen as a failure of a centralized system to control the manufacturing situation. By increasing planned lead times in response to increasing actual lead times the situation is worsened, not improved.

Progress on individual items can be scheduled and measured if a finite capacity planning system is added.

4.5 MANUFACTURING RESOURCE PLANNING (MRPII)

Poor systems breed more systems (Wight, 1981)

Growing companies tend to develop systems to solve problems as they arise. One of the problems with MRP and closed-loop MRP is that it is easy to add an extra routine or computer program to provide an extra function. In any case, these two systems do not provide any links with other areas of the business, such as the collection of payments from customers or the creation of designs. The addition of extra functions to many isolated systems can lead to systems which override one another. For example there might be a date on works or purchase orders, generated by an MRP system, which may be overridden by the dates on the urgent orders list generated by the sales department.

Wight believes that a single system and a single set of figures are vital to coordination. This is the philosophy behind MRPII. The philosophy is one of centralization and coordination, allowing considerable effort to be saved by everyone following the same system.

Traditionally, accounting figures were not based on the same information as the manufacturing figures. For example, the accounts department would have a system separate from manufacturing to calculate the value of inventory. This was because the manufacturing information was not believed to be accurate for accounting purposes – it was prepared by engineers, not accountants. Production departments would often independently calculate the value of work in progress measured in terms of standard hours.

Advanced users of closed-loop MRP systems realized that if the accuracy of the inventory records was good enough to support an MRP system then it would be good enough to be used for financial purposes. This allowed the financial and production departments to work with the same, accurate, information thus allowing them to keep the business plan up to date.

The linking of the financial and manufacturing systems is the main step in moving from closed-loop MRP to MRPII.

Wight (1981) describes the main differences between closed-loop MRP and manufacturing resource planning (MRPII):

Technically it's not much different from closed-loop MRP. It does include the financial numbers and a simulation capability.... The technical differences are small compared to the real significant difference... the way management uses the system.

MRPII systems can be used to simulate different strategies for the use of manufacturing resources. Many of them are sufficiently powerful to allow the whole of the master production schedule to be calculated and the orders put through finite capacity planning, with calculations performed on the data to determine the expected financial performance of the factory over a given period. This can be performed repeatedly to allow the best scenario to be planned, taking into account the needs of customers, the creation of a smooth cash flow, the control of inventory etc. This can be used as a 'game plan' so that all departments can pull together effectively to achieve clear aims.

MRPII is no less dependent than the other forms of MRP upon the accuracy of the data supplied.

4.6 APPLICATION OF MRP SYSTEMS

MRP systems are often used in batch or one-of-a-kind production. If a variety of items is used irregularly, MRP provides a good way of handling the large number of orders at the correct times. Where production is more repetitive, MRP can be used so that the amount of items purchased relates to the actual or forecast customer orders, rather than to arbitrary fixed stock levels. If lead times are uncertain or long, causing the MRP system to work to forecast orders with a large amount of safety stock and safety lead time, a stock control system may be just as effective. In more repetitive manufacturing systems, stock control or *kanban* methods may be used in the factory together with MRP logic for procurement of purchased parts and raw materials.

4.7 PROBLEMS ASSOCIATED WITH MRP SYSTEMS

4.7.1 DANGERS OF CLOSED-LOOP SYSTEMS

If load fluctuates then lead times will change with load (lead times grow at times of heavy load and shrink at times of light load); therefore the lead time might always be wrong because the last time a batch was made the load was different and therefore the lead time was different.

In factories that are always busy, urgent parts are expedited at the expense of delay to other work. This causes a cycle of ever increasing lead times because almost everything is late, causing the system to extend its lead time. Next time that part is made the part has a longer time allowance, but if the position is serious it may not be processed until it is urgent and arrives late again. The danger is that the lead time can be increased constantly. This increases both inventory and operational expense and therefore reduces profit. These are classic symptoms of MRP systems.

It is difficult to break this cycle. A gradual reduction of lead times may not be able to overcome the effects of queuing, and the only effect may be to make more items late.

A more radical approach to this problem may be to check whether there are particular operations which are causing delays. In particular, it must be established that extra capacity can be made available where necessary to clear the backlog. Then the drastic step of reducing the planned lead times of all items to a level which is believed to be realistic can be taken. This will have the effect of providing a slack period – if all lead times are suddenly reduced by four weeks, for example, then no more jobs will be released in the next four weeks. This should provide the opportunity to clear up some of the worst problems and get the factory in order before the load increases again. If all is successful, the factory will be able to continue under the new regime with its worst problems solved. This is a high-risk strategy, which should be considered very carefully and with more specific professional advice than can be presented here. If the problems are not solved, the situation may be much worse than before.

4.7.2 THE DANGER WITH BACKWARD SCHEDULING MRP SYSTEMS

The most efficient way to run an MRP system is with zero safety stock and zero safety lead time. This provides a theoretical ideal of no excess stock. Unfortunately, since there is no slack time this means that any delay to any component causes a delay to the customer's delivery. In an environment where each finished product contains a high number of parts it is unlikely that any product would ever be delivered on time.

In most products there tends to be a high number of lower value items and a lower number of higher value items. In order to improve the reliability of the system it is common to hold the cheaper items in stock, so that they cannot delay production, while the expensive items, which represent high inventory costs, are ordered only as required, with very little safety lead time. The smaller number of items means that these can be monitored carefully to ensure good progress.

4.7.3 LENGTH OF LEAD TIME

In attempts to make the master production schedule more achievable, safety lead time is added for each component. This tends to lead to extended lead times, particularly if the parts list is several layers deep, since the requirement date of a part on any level is offset by the lead time of the items on the levels above. The use of safety lead times and lead time rules which allow a certain number of operations per week can cause a situation in which the actual run time of the components may be only a few per cent of the total lead time, thus giving rise to a very high level of work in progress.

This can be alleviated to some extent by removing some of the arbitrary rules, and by making efforts to flatten the bill of materials, that is, to remove some of the levels. Where there are several subassemblies, their components will be at low levels of the BoM. If the subassembly levels are removed, so that all the components are seen as part of the complete assembly, all the items will be needed when the final assembly begins, which will be later than if a lead time with safety is allowed for each level.

Considering the calculator example above, the gross requirements in Box 4.6 show that all the components of the calculator are required in week 23, allowing that week for assembly; then the carton is required in week 24, allowing a week to do the packaging. If the BoM is rewritten so that the packing and assembly are all done in one operation, it will increase the time taken for the operation, but if there is a significant element of queuing time and safety lead time, these will be halved – the work need only queue once – so that one week would suffice for the one assembly operation, and all the parts could have a requirements date of week 24. In many cases, BoM structures arise from the way designs are shown on paper by designers, which is related to clarity of understanding and should not be taken as an instruction on the amount of assembly work to be done in one operation.

4.7.4 STANDARD ORDER QUANTITY

If the bill of materials has many levels, and items at each level have standard ordering quantities, it is possible for small orders to generate large amounts of inventory.

Referring once more to the calculator example, if there were a standard order quantity for lower case assemblies of 100 units, this quantity would be ordered even though only 65 are actually required (week 22). In some MRP systems, this would then result in a requirement for 100 lower case mouldings and 100 battery covers, even if there were enough in stock to meet the actual requirement of 65. The production manager has to be very vigilant to identify orders which are generated in this way.

This effect can be caused by the use of economic ordering quantities, and can be particularly troublesome in cases where there are many levels in the bill of material, with standard quantities at each level.

SUMMARY

MRP seems ideal. The system has all the information about all the jobs that need to be performed and the dates by which they are required. Therefore (theoretically at least) it is possible for a computer to calculate the optimum schedule. In practice it is unlikely that the all the estimated timings would be correct. The accumulation of the differences between actual and estimated times causes problems with the schedule. To reduce this problem, safety lead time is increased. This approach can lead to long lead times and high levels of work in progress.

MRP is the most widely used CAPM system and is used in a large range of environments. The main drawbacks with MRP are that it requires high data accuracy in order to be successful and can lead to extended lead times and high inventory levels. It is easy for MRP to get out of control.

QUESTIONS FOR DISCUSSION

1. What is the basic function of a standard MRP system?
2. What is a bill of materials?
3. What is the difference between gross requirements and net requirements?
4. What is the effect of an inaccuracy in the bill of materials file?
5. What is the effect of an inaccuracy in the inventory file?
6. What precautions can be taken to avoid problems in MRP systems?
7. How is closed-loop MRP different from standard MRP?
8. Does closing the loop create a better system or a system which is harder to control?
9. In what ways are the three main variations of MRP similar to one another?
10. MRP systems allow the computer to take many decisions in the management of production. Does this save effort or reduce control?
11. Referring to the calculator example described in the text, determine the order schedule for the other components. The further information required is shown in Box 4.11.

Box 4.11 Item master file information and inventory status information for all items.

Item	Available stock	Lead time (weeks)	Make/Buy
Adhesive badge	100	3	B
Battery cover	20	2	B
Calculator complete	0	1	M
Calculator packed	0	1	M
Carton	200	3	B
Circuit board assembly	10	4	B
Insert	50	2	B
Instruction leaflet	60	3	B
Keypad	30	4	B
Lower case assembly	35	1	M
Lower case moulding	15	2	B
Screw	60	1	B
Transparency	30	1	B
Upper case assembly	20	1	M
Upper case moulding	0	2	B

12. The plans as calculated have operated until week 21. It is now discovered that there are only 25 lower case assemblies in stock (not 35). What effect will this have on deliveries to your customers, and what changes will you make to your planned orders?

REFERENCES

Browne, J., Harhen, J. and Shivnan, J. (1988) *Production Management Systems – A CIM Perspective*, Addison-Wesley, Wokingham.

Callerman, T. and Heyl, J. (1986) A model for material requirements planning implementation, *International Journal of Operations and Production Management*, **6**(5).

Wight, O.W. (1981) *Manufacturing Resource Planning: MRP II – Unlocking America's Productivity Potential*, Oliver Wight Ltd Publications, Essex Junction VT, USA.

Just-In-Time 5

5.1 INTRODUCTION

Western interest in 'Just-In-Time' manufacturing, or 'JIT', grew through-out the 1980s as both manufacturing companies and academic researchers tried to come to terms with the high penetration of markets in the West by Japanese companies. Originally seen by many as being 'the way the Japanese work', JIT can now be seen as a philosophy of good business practice which was influenced by Taiichi Ohno and Shigeo Shingo at Toyota and widely reported in the work of, amongst others, Richard Schonberger.

Just-In-Time is a philosophy which aims to eliminate waste. It does this through reducing inventory and increasing throughput and quality.

Ohno (1988) uses the following definition of Just-In-Time:

> Just-in-time means that, in a flow process, the right parts needed in assembly reach the assembly line at the time they are needed and only in the amount needed. A company establishing this flow throughout can approach zero inventory.

Schonberger (1982) has described JIT as the ability

> to produce and deliver finished goods just in time to be sold, sub-assemblies just in time to be assembled into finished goods and purchased materials just in time to be transformed into fabricated parts.

The fundamental concept of JIT is that it is good practice to manufacture what the customer requires at the time it is required, and nothing else, rather than to tie up working capital and space in inventory.

JIT is commonly associated with *kanban*. The difference is that Just-In-Time is a philosophy for the reduction of waste, whereas *kanban* is a control mechanism for controlling production on the shop floor, with low levels of inventory. *Kanban* will be described later in this chapter.

5.2 THE PHILOSOPHY OF JUST-IN-TIME

In essence JIT means that we make 'what we want when we want it'. If a customer buys products one at a time, it makes sense for the factory to produce them one at a time, since to make more would raise the level of inventory. It is good business sense to make what is needed rather than to invest in stocks of finished goods. This can be seen clearly by using the indicators described in Chapter 2. Increasing inventory ties up capital, incurs handling costs and increases operational expense, and thus reduces net profitability.

5.2.1 JIT AND BATCH SIZES

Sandras (1989) claims that the way to move towards Just-In-Time is 'to learn how to economically manufacture "One Less at a Time"'. As constraints are exposed these should be tackled, and then the process should continue. If a reduction in batch sizes causes a capacity problem because of the extra time spent on set-ups, then work should be done to reduce the set-ups to make the process economical. To reach a batch size of one is not the only goal, but once this has been achieved attention is focused on defects and then on any remaining non-value-added activity which can be reduced or eliminated.

In the traditional Western style of manufacturing, where batch sizes are kept large to amortize set-up times, inventory levels would be high and the load in the factory would be lumpy. Batches of items are processed and moved together, so that each item waits for the whole batch to be finished at each operation before they all move on together to the next operation.

In a JIT factory, where set-up times are reduced to allow very small batches, the inventory would be much lower. Consider the following example of large and small batches. Demand is the same in each case, for products A, B and C in the ratio 3:2:1, and we assume that each item takes a day to manufacture.

In the first case, each item is produced on its own, in a batch of one. This means that whether product A, B or C is ordered, it can be made tomorrow.

Production ABABACABABACABABAC   Total production 9A, 6B, 3C
Inventory 1 1 1 1 1 1 1 1 1 1 1 1 1 1 1 1 1 1   Average inventory 1 item

In the second case, products A, B and C are manufactured in batch sizes which have been set at 9, 6 and 3 respectively. This means that over the 18 days of production shown, only three set-up changes will be needed for each operation.

Production AAAAAAAAABBBBBBCCC   Total production 9A, 6B, 3C
Inventory 1 2 3 4 5 6 7 8 9 1 2 3 4 5 6 1 2 3   Average inventory 4 items

If the orders are combined into large batches, the customer for item A will have to wait up to nine days. The alternative to this is to hold a further nine items as finished goods stock, which doubles the inventory. The only advantage to producing in large batches is the reduction in the number of set-ups required. However, if set-ups can be done quickly and cheaply, this advantage is much less important, and the JIT approach of manufacturing in line with the quantities ordered by the customer is much more economical.

If products can be manufactured economically in batches of one, there will be no excess work in progress and a minimum stock-holding cost.

Where items are always purchased in larger quantities, such as nails or bricks, the 'batch of one' should be interpreted as meaning 'one of the things as the customer orders it', such as a box of nails or a pallet of bricks. By moving towards the batch of one, the company begins to treat the manufacture of discrete items as a flow of products, smoothing out variations in demand and reducing their effects.

Traditionally, it has been common to calculate batch sizes using the notion of economic batch quantity, balancing inventory costs against lost production due to set-up times. This method assumes that the set-up time is a 'given' and therefore matches uneconomic long set-up times with uneconomic long batches. Shingo (1985) has drawn attention to this and developed the SMED (single minute exchange of dies) system, a simple approach to the reduction of set-up times. Although originally developed for the rapid exchange of press tools in automotive plants, the methodology applies to all machine tools and equipment.

The SMED methodology first looks at the elemental activities involved in the set-up and classifies them as Internal or External. Internal set-up operations must be done on the machine and require the machine to be stopped, such as the loading of jigs to the worktable. External operations can be done off the machine while it is still running the previous job, such as finding and assembling tools. All internal operations must be examined to see whether they can be converted to external operations. This can mean, for instance, providing location devices and presetters to allow all adjustments to be performed away from the machine. All remaining internal operations must then be examined to see how they can be improved. For instance, quick change devices such as those already commonly used for the location of components in fixtures can be used for the location of jigs and fixtures in machines.

Shingo's approach, which consists of a few engineering tips and a serious attitude to a problem that has often been ignored, has resulted in some enormous time savings, such the changeover on a six-axis boring machine being reduced from three days to 2 minutes 40 seconds. This approach means that small batches or single items can be produced with enormous savings in inventory.

JIT's small batches have other benefits apart from inventory reduction, such as the reduced number of items that are lost through incorrect processing or design obsolescence. Lead time is also affected. Reducing the batch size reduces the lead time, especially where there is little or no queuing between operations, such as in applications of group technology (Chapter 7).

5.2.2 JIT AND WASTE

In a JIT environment, 'waste' is defined as any activity that does not add value for the customer. It includes the use of resources in excess of the theoretical minimum required in terms of staffing levels, equipment, time, space or energy. Waste can be excess inventory, set-up times, inspection, material movement, transactions or defects. Any process that does not actively add value is waste and should be minimized and eliminated. Shingo (1985) has defined the types of waste as:

- *Waste of over-production*
  Over-production, or production of items too early, which causes unnecessary inventory. The classic example of this is batches larger than the amount ordered by any customer.

- *Waste of waiting*
  This refers to operators. Operators should always be occupied doing something useful. This may include preparing for the next job, maintaining equipment or cleaning up. If machine operators are idle while the machine operates, they should operate more than one machine.

- *Waste of transportation*
  Poorly designed layout causes unnecessary transportation. This adds no extra value to the product and therefore is wasteful. In many factories, particularly those with a high product variety, there is often no clear route through the factory taken by items in general. An often quoted example is the missile factory where it was found that the product travelled further during manufacture than it would do when fired.

- *Waste of stocks*
  Stock itself is wasteful because it ties up capital and also because it needs to be transported, stored and found when needed. Many businesses have invested huge sums in sophisticated systems such

as automated warehouses with expensive computer systems and staff employed to put away and find stored items. All of these represent waste.

- *Waste of worker motion*
  Unnecessary fetching and carrying of materials or components adds no value, therefore it is wasteful. In many factories, production workers waste time looking for material, queuing at tool store issue counters and carrying things from place to place.

- *Waste of making defects*
  Defective items may be scrapped, in which case both all the work done and the material used were wasted. Rework to correct defective items, where it is possible, is classed as a waste. Defective parts may also cause damage when assembled, or may cause an assembled product to fail. The inspection and testing activities which are needed to identify defective items are also seen as waste, since they would not be needed if ways could be found to avoid making defective items in the first place.

- *Waste of processing* (when the process should not be used)
  This refers to using a machine which is more than capable of doing the job, such as a machine with excess power or accuracy. This means that the opportunity of using the machine for something for which it adds more value may have been wasted.

- *Waste of operation*
  It is wasteful to use poor methods which can cause quality problems, damage to parts at later stages and unnecessary rework.

Taiichi Ohno (1988) in particular stresses that JIT is about making more with the same or fewer resources. If a production line of ten people improves production by 25% then the same volume can now be produced by eight. Ohno would suggest that the extra two people can be used in other ways, since the efficient JIT business will hopefully prosper and expand, so that it will be able to sell 25% more. If this is not the case, Ohno would suggest that they should be made redundant.

The fear of redundancy is a problem associated with JIT in the minds of many people. A manager who is convinced of the growth potential of the business should ensure that staff are retained, so that skills are not lost which will be needed as the company expands. This is particularly important when large productivity gains are made, since it is often convenient to allow 'natural wastage' to reduce excess staff numbers. The staff who tend to be attracted by opportunities such as early retirement are often exactly the ones who it is vital to retain. A manager who is not convinced of the potential of the business should not be allowed to destroy the goodwill of the employees and let essential skills evaporate.

The question of employment is difficult to deal with. It is clear that a company which operates inefficiently will not survive for long in a competitive market. Competitive companies, on the other hand, are likely to be able to grow and, hopefully, the more competitive they are the more they will grow. This should mean that improvements in competitiveness would be welcomed. Unfortunately, the threat of redundancy has often acted to oppose initiatives to improve manufacturing companies. One reason for this is that those employees who fear losing their jobs do not expect to be able to find new ones. To remove this fear requires a general and continued improvement in the competitiveness of industry in the country. This means that, where possible, all companies must improve. It also raises political questions with respect to the conditions which are necessary for industry to grow. In the case of the individual company, every attempt must be made to use all the resources available, which includes the skills and experience of all employees. There are great dangers for the loss of flexibility if companies are slimmed down so that all apparently redundant skills are lost. A policy of reducing cost leads to the loss of the business, since no business can operate with zero costs. Nevertheless, if employees are genuinely and permanently redundant, it is better to let them go than to deliberately waste their skills and carry excess costs.

The question that should be posed when implementing a JIT philosophy is, 'how can the processes be changed to eliminate the need for non-value-added activity or inventory?'. There are a number of key components of the JIT philosophy which need to be examined in detail.

5.2.3 JIT AND TOTAL QUALITY

Van Loon (1990) defines total quality management (TQM):

> TQM aims to achieve step by step continuous improvements in the organisation in order to establish a competitive edge and strengthen market position. TQM aims at customer satisfaction through customer orientation, always ensuring that... improvements respond to market demands.

Companies attempting to use JIT techniques are removing the comfortable slack (safety stock and safety lead time) which allow production to continue on most work centres most of the time. The Just-

In-Time philosophy helps to overcome quality problems, line imbalances and machine breakdowns. Safety buffers were put in place for good reasons, so in removing them precautions must be taken to avoid disastrous results. Sandras (1989) considers that

> Just-in-Time and Total Quality Control are two sides of the same coin... Just-in-Time is going to force you to practice Total Quality Control.

If all the wastes are to be removed, a tool is required to deal with the problems which will appear when the slack is removed. Total quality approaches provide techniques to ensure conformance to requirements and to eliminate variations. One of the more obvious aspects of TQ is to make individuals or groups responsible for the quality of their own work, rather than allowing inspectors to weed out defective items.

Schonberger (1982) states:

> It is not enough however to simply announce a transfer of responsibility for quality from inspectors to operators. Training is needed in the principles of Statistical Process Control (SPC) to allow operators to correct problems before defects are manufactured. They can also be trained to perform simple maintenance and adjustments on their equipment. The most important factor is to ensure that the members of the company can appreciate the effect of quality on the organisation.

Equipment must be both capable and available. Capable machines are those whose output tolerances are statistically acceptable with reference to the design tolerance of the item produced. Availability refers to the amount of time the machine is able to be used for production. This is kept high by ensuring that planned preventative maintenance (PPM) is carried out on a regular basis, to ensure that unpredicted breakdowns, which will significantly disrupt a JIT company, are kept to an absolute minimum.

The quality of supplies is also important to JIT – this means both the correctness of the products supplied and also their delivery on time. Either defective supplies or late supplies will cause a standstill in production. Many JIT companies are putting great stress on the quality of supplies and are engaged in supplier development programs. This means educating the suppliers about JIT and total quality, especially SPC, and in many cases sending out methods engineers to help suppliers to develop their own methods. At the same time this means developing a trusting relationship with suppliers, and usually only allowing one source for each item, with long-term ordering.

Another element of TQ which is very relevant to JIT is the establishment of a system of performance measures which can show the company how its improvements are working. This is important to retain the commitment of both management and employees to continuous improvement.

5.2.4 WHAT IS QUALITY?

It is important to realize when discussing quality that we are not talking about the 'goodness' of the product. Crosby (1979) defines quality as

conformance to customer requirements.

Deming (1982) goes a small stage further and suggests that quality is

meeting or exceeding the customers' requirements.

As customers' requirements and expectations increase, the company is already prepared and has an advantage over the competition who will need to improve their quality. The Deming approach is one of disconnecting the mental link which implies that higher quality means higher cost. Combined with JIT, quality means operating the business in a way which allows customer requirements to be met in the most effective and least wasteful manner, while being prepared to meet ever-higher demands.

5.3 JUST-IN-TIME PROCUREMENT

The 'Just-In-Time' approach extends to procurement of raw materials and purchased parts; for example, Ohno (1988) explains that Toyota have deliveries to their plants in Japan several times per day of batches which allow only a few hours' production.

Objections to this are usually based on the cost of producing many purchase orders and receiving many shipments. More frequent deliveries mean that less space is required for storage and that deliveries can be made direct to the production line. This is particularly important for Toyota in Japan, where storage space is extremely expensive and purchased items, such as car seats, are very bulky to store. This decreases administration and internal costs associated with storage, since no procedures are needed to place stock in stores, record its location or, later, to retrieve it. It also allows the space to be used for other purposes.

In high-volume production such as the above Toyota example, JIT procurement has been shown to be feasible. Companies like Toyota have very great buying power, because of high volumes in the car industry. This buying power allows the company to dictate terms of delivery to its suppliers. In contrast, high-variety companies tend to buy relatively small volumes from a large number of suppliers. It is often argued that this means they have too little influence over any one supplier to persuade them to provide JIT deliveries. In such cases companies must work towards reducing the number of suppliers they have and concentrate the JIT procurement efforts on the highest value components. Rather than giving up JIT because direct deliveries to the factory are impractical, companies can benefit from more frequent delivery of smaller numbers of items – if deliveries are made twice as often, inventory should be approximately halved. This is a step towards JIT.

Another problem which may arise from JIT implementations is that stock is pushed onto suppliers, thus increasing the cost, quality control and volumes of paperwork. These are discussed later in this chapter.

5.4 JUST-IN-TIME SHOP-FLOOR CONTROL

It is important to remember that Just-In-Time is a philosophy for reducing waste and can be applied with any control system. JIT can be said to operate in other control systems where inventory, especially safety stock, is kept very low, lead times are kept short, and manufacturing batch sizes reflect the numbers ordered by customers rather than predetermined batch sizes. These are some of the aspects of good production management, which can be achieved by various approaches. Nevertheless, the most commonly discussed form of Just-In-Time control is *kanban*.

5.4.1 THE RELATIONSHIP BETWEEN JIT AND *KANBAN*

There is sometimes confusion about the difference between JIT and *kanban*. JIT is a philosophy, which dictates that inventory should be reduced and waste eliminated, while *kanban* is the shop-floor control system developed by Toyota Motor Company as part of their implementation of JIT.

The confusion is probably because the company that brought JIT to the attention of the world is Toyota and therefore the Toyota method is most widely discussed. At Toyota, Ohno believed *kanban* to be more than a Just-In-Time production system. It includes waste elimination at all levels, multi-machine handling by a single worker, 'autonomation' (machines that automatically stop production if errors occur), 'Baka Yoke' (fail-safe mechanisms which prevent mistakes), preventative maintenance and 'The Single Minute Exchange of Dies' (SMED) system, a handbook of engineering techniques for set-up time reduction. The SMED system is described fully in Shingo (1985).

Kanban is the Japanese word for card, sign or signal. In this context it means the cards that signal or regulate production. It does not have to be an actual card, but can be any signal. For example, in one Japanese factory they use a golf ball which is rolled down a tube to the appropriate workstation. Often containers are used instead of cards, and *work centres only produce if empty containers are sent back for filling*. The key to understanding the *kanban* system is that no work is done on any item unless that item is required by a downstream workstation or a customer. Thus the system ensures that no work is done and no material is used unless there is an actual requirement for it.

The amount of work in progress inventory (WIP) is regulated by the number of cards and the number of items ordered by each card. The number of cards or containers can then be reduced, or their size may be reduced, in order to reduce the amount of capital tied up in WIP at any time. The process of reducing the amount of WIP continues until it is discovered that a work order cannot be fulfilled quickly enough. This may be when a work centre receives more cards while it is still busy with a previous one, showing that its production capacity is too low to meet the current value of varying demand, whereas previously there was enough inventory to cover such a peak. At this point production is stopped and attention is paid to why this happened. It may be necessary to reduce the set-up on a work centre by investing in some tools, thus reducing the batch size which can be manufactured economically. *Kanban* thus provides a way of controlling production which has the built-in facility of allowing inventory to be measured at a glance and gradually reduced, thus providing a means of continual improvement.

However, inventory reduction and continual improvement are not restricted to *kanban* systems, and if *kanban* were to be implemented on its own they would not necessarily follow. The company must also be in the right mood to accept continual change to working arrangements, be conscious of the need to improve quality, be prepared to work cooperatively as a team etc. The implementation of a *kanban* system would not automatically give a company the productivity figures of Toyota. It is important not to be misled into thinking that the good results of a productive company are due to the visible differences, when the invisible features may be far more important. Shingo (1989) puts it thus:

Some people imagine that Toyota has put on a smart new set of clothes, the kanban system, so they go out and purchase the same outfit and try it on. They quickly discover that they are much too fat to wear it! They must eliminate waste and make fundamental improvements in their production system before techniques like kanban can be of any help. The Toyota production system is 80 percent waste elimination, 15 percent production system and only 5 percent kanban.

5.4.2 TYPES OF *KANBAN* CONTROL SYSTEM

There are several forms of *kanban* control system, suitable for different environments.

Single-card *kanban*

In a single-card *kanban* system there is only one type of *kanban*, which is used to request the manufacture and movement of the component. Figure 5.1(a) shows the operation of a single-card *kanban* system in the manufacture of the same product as in the previous chapter, the calculator.

Firstly, it will be noticed that the factory or part of the factory used to make the calculator is laid out in specific areas. There are two workstations dedicated to the production of each subassembly, one performing the assembly and test of the calculator and one packing the products ready for sale.

At each workstation is a small buffer stock of the item made there, and there is a buffer of items delivered by suppliers. Each item in each of the buffer stocks has with it its *kanban* card. Each card is clearly identified with the name and part number for the item it controls, and the number of items which it authorizes. In this example, each card authorizes only one of its item, except for the cards controlling screws, which each authorize three screws.

When all the items are in place, there are no cards which do not have goods with them and therefore nothing happens. This is the position shown in Fig. 5.1(a), which shows all the buffers full and no instructions for production. When a finished product is sold to a customer, it leaves a card without an item, as shown in Fig. 5.1(b). The card which was with the product is returned to the workstation concerned, as the instruction to make a new product to replace the one that has been sold. In the simplest *kanban* systems, the card and item would not be far away from their producing workstation, but the logic is still that the workstation receives the signal to produce a new item.

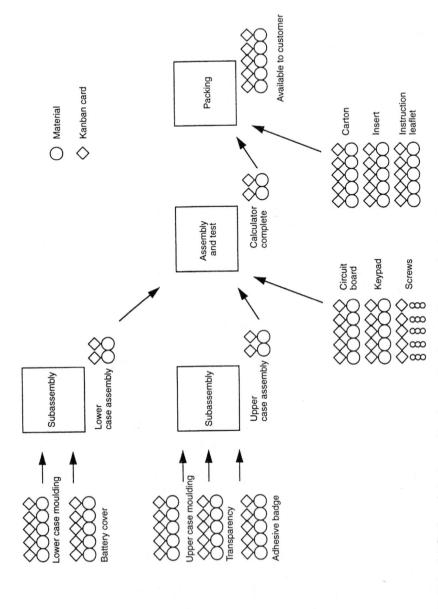

Figure 5.1 (a) Outline of single-card *kanban* system for calculator production.

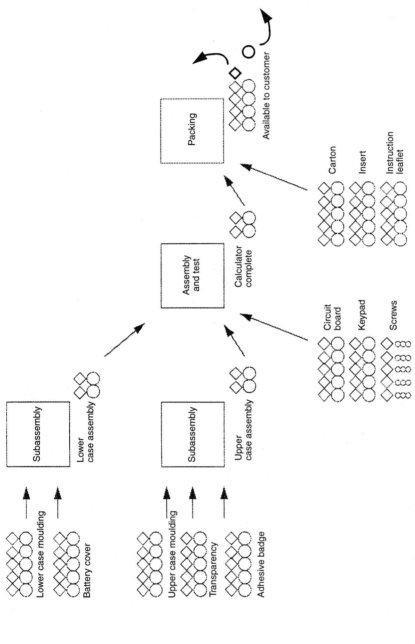

Figure 5.1 (b) Product sold to customer, leaving *kanban* card.

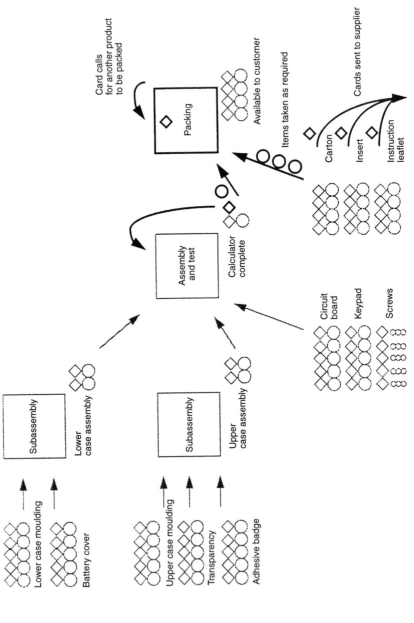

Figure 5.1 (c) Packing station receives signal and packs a new product.

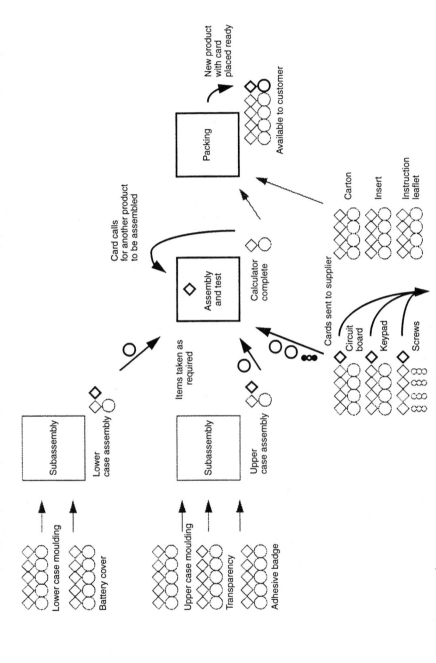

Figure 5.1 (d) Assembly station receives signal and assembles a new product.

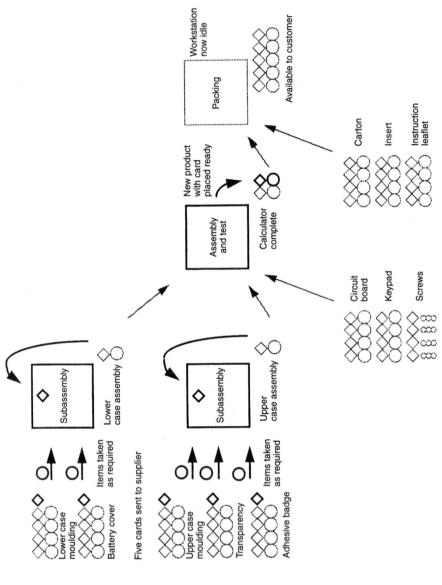

Figure 5.1 (e) Subassembly stations receive signal and assemble new subassemblies.

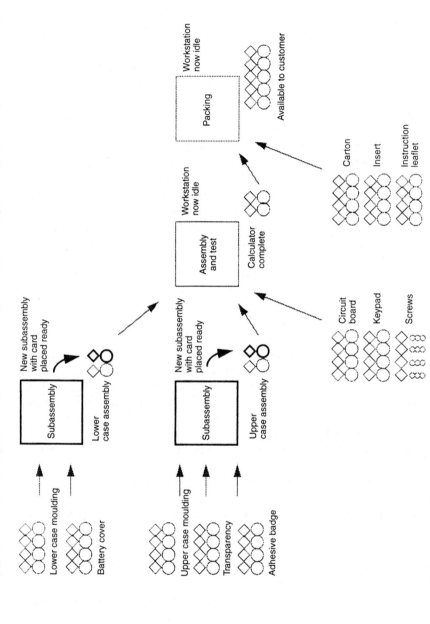

Figure 5.1 (f) Subassembly stations complete work.

On receipt of the card, or on seeing that the card is now without its item, the operators of the workstation set about the work of producing the new item. As shown in Fig. 5.1(c), this is the packing station. The packing materials (carton, insert and instruction leaflet) are taken from their respective buffer stores and used to make the packed product. The withdrawal of the three packing items and the product leaves four cards without items. These act as the signals to the supplier to bring in some more packing materials, and to the assembly and test workstation to assemble and test another calculator.

When it is finished, the packed item is placed in its buffer store with the *kanban* card which authorized it, as shown in Fig. 5.1(d). As there are no more 'empty' cards at the packing station, work stops. The assembly and test workstation now has an 'empty' *kanban* card, so the operator takes the lower and upper case assemblies, the circuit board, keypad and screws and assembles and tests another calculator. This leaves empty cards at each of the subassembly stations calling for work to be done, and at the component store location calling for the replacement of the purchased items.

The tested calculator is placed with its card in the buffer stock location, as shown in Fig. 5.1(e). At each of the subassembly stations, work begins, because each station has a *kanban*. These both draw materials from buffer stocks of purchased items, leaving empty *kanban* cards for the supplier to fulfil.

The completed subassemblies are placed back in their buffer locations (Fig. 5.1(f)), and unless more orders are received all the workstations will now become idle. While it is not desirable for production to stop, the *kanban* mechanism ensures that production only meets customer requirements as they arrive, and no excess stock is produced beyond that which is placed in the buffers.

From the example it can be seen that the buffers of purchased items have all been reduced by one (since one product was purchased). The buffers of purchased items must cover the expected production between deliveries from the supplier.

Suppliers may act as the first station on the line, supplying parts in direct response to *kanbans*. However, if the supplier chooses to make less frequent deliveries of larger amounts, this can be accommodated by allowing the supplier to collect *kanbans* until the supplier chooses to make a delivery. This can operate on the basis of two rules:

- that the items supplied are the property of the supplier and are not bought and paid for until they are taken from the buffer and used in production, which means that the manufacturing company's stock-holding cost is the same as if the items were only delivered Just-In-Time

- that the supplier guarantees to maintain the buffer store at an agreed level.

Another alternative is to order from the supplier all the parts which will be needed to replenish those which will be used by orders that have been received. This means that the supplier is told about requirements as soon as the customer order is received, rather than when the signal ripples back through the *kanban* line. This means that purchasing can be directly linked to sales, and is often performed using a single-level bill of materials which shows only the purchased items, as described later in this chapter.

If demand is continuous, all the workstations will operate for most of the time. Where the time taken to do the work needed at each station is different, the number of people working at each station can be adjusted until a balance is found. With a reasonably stable level of demand, the *kanban* mechanism is not easy to see, since all the workstations are in continuous operation. The *kanbans* provide a link between the workstations to ensure that all operate at the same pace. Any workstation which cannot keep up with the pace will be identified easily, as will any problem which means that production stops.

In the *kanban* system as operated at Toyota any operating problem is treated as an opportunity to improve. As long as no-one covers up quality problems, the effect of a workstation being unable to produce good parts will quickly become obvious as the supply of those parts to workstations downstream dries up. When the line stops, the operators on the line should help solve the problem. A system of warning lights known as the 'Andon' system is used in some implementations. Different coloured lights can be illuminated by operators to ask for assistance if they experience delays, quality problems, machine problems etc. This further accentuates the *kanban* system's principles of maintaining output and drawing attention to and solving problems.

The *kanban* system can benefit from multi-skilled workers, even though each workstation has a specific and fairly narrow role. If the operators have a range of skills they can move from station to station either to balance the line or to help correct a defect. In a variation of the single-card system, each card is placed on a central display board, to be picked up by the next operator to become available. The operator completes the operation and delivers the item to its correct location and then returns to collect the next *kanban*. This system is able to balance itself more easily and allows the same workstations to be used for several products, without each requiring its own separate part of the factory.

If the *kanban* production area is arranged in a form which corresponds to the flow of materials, the single-card system can function without the cards moving, if the work and cards are stored on a rack between workstations. The *kanban* card can then stay on the shelf, where it can be seen if it is without the item it controls. If the number of items at the workstation is not too large, the shelf itself can be the *kanban* – when there is a space for a certain item, that is the signal for the item to be produced. A drawback of this system is that when there are several item demands it is not obvious which one has been waiting the longest.

Standard work-holding containers are often used for *kanbans*. Each container represents a certain quantity of a particular item, and the containers circulate acting as signals – an empty container is the signal to produce the item it holds. This is common in industries such as the car industry, where items are not usually produced singly but in large quantities. The manufacturers of work-holding boxes make this simple by providing standard boxes with a slot to accept a *kanban* card, which gives each box its required identification. Where each container always holds the same item, it is possible to design containers to suit a particular item to prevent handling damage.

The single-card *kanban* system is most suited to continuous production of a narrow range of items – at least by comparison with MRP (Chapter 4). Many *kanban* implementations make use of master production scheduling to level out the peaks of demand so that the *kanban* system is able to operate without too much disturbance. However, if the type of business has too much variety, a high level of forecasting will be required to anticipate the peaks. In these circumstances, it is quite possible for a *kanban* system to produce something which is not needed.

A wider range of items can make shop floor layout and control more difficult because of the number of different items and cards which must be managed. When the volume of each item produced is low, the buffer stock between workstations means that inventory level could be quite high and the *kanban* system may show a low return on investment in comparison with MRP, where the theoretical minimum stock level is zero.

Dual-card *kanban*

A dual-card *kanban* system is shown in Fig. 5.2. This may be needed where lot sizes or components are large and there are many different varieties of item in production, because it allows the inclusion of off-line storage. Each workstation has its own input store for materials to be used, as well as its own output store of finished items. These stores are not necessarily located with the workstations, although it would increase wasteful transportation if they were far away.

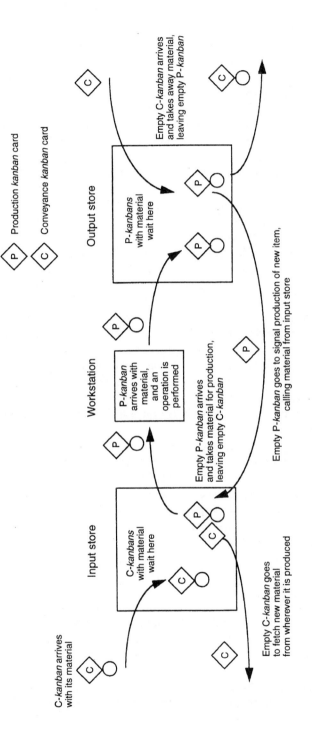

Material

P Production *kanban* card

C Conveyance *kanban* card

Input store

C-*kanbans* with material wait here

Workstation

P-*kanban* arrives with material, and an operation is performed

Output store

P-*kanbans* with material wait here

C-*kanban* arrives with its material

Empty C-*kanban* goes to fetch new material from wherever it is produced

Empty P-*kanban* arrives and takes material for production, leaving empty C-*kanban*

Empty P-*kanban* goes to signal production of new item, calling material from input store

Empty C-*kanban* arrives and takes away material, leaving empty P-*kanban*

Figure 5.2 Outline of dual-card *kanban* system.

Two types of *kanban* card are used. A C-*kanban* (conveyance) is used to allow the transfer of parts from the output store of one workstation to the input store of the next. This is also referred to as a withdrawal *kanban*, since it allows items to be withdrawn from stock. A P-*kanban* (production) is used to regulate production within the work centre, that is, to replenish the output stores when a finished item is taken out.

A conveyance *kanban* orders material to be moved from the output store of the workstation where it has been made to the input store of the workstation where it will be used, or to the customer. The C-*kanban* arrives at the output store and is attached to the required item. The item is moved, leaving behind the production *kanban*, which has been with the item since it was produced.

The production *kanban* provides the authorization to draw the required material(s) for production of the item from the input stores to the workstation, and provides the signal to the workstation to produce the item needed. The finished item is placed with the P-*kanban* in the output store.

At the input store, when the material is taken for manufacture, as authorized by the P-*kanban*, it will leave behind the conveyance *kanban* which has been with the item since it was conveyed from the output store of the workstation where it was made. This C-*kanban* is now sent back to authorize the conveyance of another item from the appropriate output store.

The chain of signals passes backwards along the line, resulting in the demand for items from suppliers.

Variations on dual-card *kanban* are possible. The output store of one workstation can be combined with the input store of the following one, but this means that items are only able to flow along one route. It is also possible to combine all the stores into a single store, which allows different products to follow a wide variety of different routes.

Single and dual *kanban* systems are usually used in high-volume production applications where a small group of items are required repeatedly.

Brand name *kanban*

Brand name *kanban* is used in systems where there is a constant demand for a series of items, but allows the items to be customized in some minor way, such as a different brand name. Requests for production of a particular brand name are issued using *kanban* cards and orders are satisfied as usual. The difference from single-card *kanban* is that the card specifies which brand should be manufactured. This can only operate, however, when each brand is interchangeable with other components and can come from the same source. For example, an international personal computer manufacturer may use brand name *kanbans* to issue orders for keyboards, where the *kanban* will specify the required key layout for a particular language or country.

If the variation is higher or if the number of *kanbans* needs to be reduced below the level of the number of brands, then generic *kanban* may be considered.

Generic *kanban*

In generic *kanban* the card or signal does not specify the actual part to be made but allows production of the next part of a particular type in the order schedule. For example, it allows the production of a car radio, but the specific model type must be found from the schedule. This ensures that production is always for an order, not automatically to replenish stock. If a *kanban* appears for a generic part type, but there is no specific requirement for a part, no parts are produced. To manufacture parts before there is a definite demand for them would be wasteful.

In many ways generic *kanban* is similar to synchro-MRP (see below) because production of a variety of parts is possible. Generic *kanban* can handle higher variation parts which only need to be capable of being processed by the same machines or in the same manufacturing cell. It is used in situations where low volumes of any given variety are required. Synchro-MRP is used where there is repeatability of demand but products are being made in low batch sizes (including a batch size of one).

Synchro-MRP

A system named synchro-MRP used at the Yamaha motor company has been described by Bullinger *et al.* (1986), Schonberger (1982) and Hall (1981). This is essentially a *kanban* system, but instead of following the simple logic of the ordinary *kanban* system, which means that cards are acted upon in the order they arrive, the operators ensure that there is a works order that will use the goods before processing them. This means that the level of stock of each item is not automatically replenished, but production is held back until the item will actually be taken and used. MRP is used to procure the materials from suppliers.

This system can only be used when demand is very regular and each item can always be made in the time allowed between the *kanban* being released and the item being used.

Ohno (1988) describes how the system at Toyota achieves the same WIP control. When the car being assembled reaches a certain position on the production line, the operators assembling the stereo system know that one is required. It is then assembled ready to be installed as the car proceeds. The car itself acts as the *kanban*.

This system is only practical in a repetitive manufacturing environment and achieves the same as generic *kanban* systems for medium variety manufacture, that is, that there is never any WIP that is not actually needed.

Many *kanban* systems have a computer planning tool to explode the bill of materials and purchase materials. Synchro-MRP is only different because it checks that a part is needed before fulfilling a *kanban* production request. In this way it eliminates the need to have temporary *kanban* stores between work centres.

Figure 5.3 illustrates the types of *kanban* system and their flexibility.

Figure 5.3 Various types of *kanban* system and their flexibility.

5.4.3   SINGLE-LEVEL BILL OF MATERIALS FOR MRP PROCUREMENT IN JIT

When a *kanban* system is used for shop-floor control, it is sometimes possible to arrange for suppliers to act as an extension to the system by operating in response to *kanban* cards, as was described earlier. Where this is not possible, a bill of materials can be used with a simplified MRP system. Whenever a product is ordered, the *kanban* system will require more of each purchased item in order to replenish the buffer stores. These items are simply ordered by reference to the single-level bill of materials as shown in Box 5.1. This is also called a 'purchasing BoM'. The issue of purchase orders can be triggered either by the completion of a product or from the master production schedule.

5.4.4 THE APPLICATION OF JIT CONTROL SYSTEMS

It is very difficult to set up *kanban* flow arrangements unless there is a certain amount of predictable, repetitive demand. *Kanban* is associated almost entirely with repetitive manufacture. Voss and Harrison (1987) state:

... The nature of jobbing and projects makes them unsuitable for the flow elements that are the core of JIT.

Box 5.1   Simplified bill of materials for purchasing.

Calculator packed
　　Lower case moulding 1
　　Battery cover 1
　　Upper case moulding 1
　　Transparency 1
　　Adhesive badge 1
　　Circuit board assembly 1
　　Keypad 1
　　Screws 3
　　Carton 1
　　Insert 1
　　Instruction leaflet 1

In contrast, Sandras (1989) believes there is greater benefit from moving to Just-In-Time in job shop applications than in repetitive manufacture because the former contain more waste. However, it is difficult to conceive of any kind of *kanban* system which would be able to operate in a low-volume, high-variety factory where every product is different from previous ones. A rough guide would be that *kanban* can be used in make-to-stock, make-to-order and assemble-to-order businesses, where product variation comes from a range of possible components. Where products are actually designed or engineered to each customer's order, the differences in the products are likely to be so great that a practicable *kanban* system could not be designed to suit all the possibilities.

It is possible to use *kanban* in job shops where families of similar parts requiring similar processes can be identified. If the variation is very high the problem is one of loading the manufacturing facility to avoid impossible peaks of load and slack periods as the product mix changes.

5.5   ARGUMENTS AGAINST JIT

JIT is not a single objective, but rather a journey of continual improvement with a clear direction (ever decreasing waste). The implementation of JIT is not easy and involves a lot of work both with suppliers and in-house. Many problems need to be solved and many companies have found solutions to them. Despite the great successes of JIT it is often easier for production managers to proclaim that 'JIT would not work here' than it is to face the problems and find solutions.

5.5.1 HIGH COSTS OF MULTIPLE JIT DELIVERIES

Burbidge (1987) proposes a 'call off' system whereby the schedules of requirements are given to a supplier for up to a year in advance. For example a company might make an order for 200 of a particular component to be delivered in weekly lots. The supplier then supplies each week the number 'called off'. This has the advantages that only a single order is required, demand is smoother, cash flow is improved, inventory is reduced and a long-term partnership develops.

This, however, does not solve the additional cost associated with extra deliveries. Tom Peters (1987) explains how Hewlett-Packard solved this particular problem. They asked their suppliers for quotes for the quantities required both including and excluding delivery. They then obtained a quote from a haulage firm to deliver from all the supplier companies twice per day. Together with the costs of the components this was considerably less than the delivered costs and made large numbers of small deliveries economic. Burbidge (1986) proposes the same, and an alternative solution to the problem of multiple deliveries. He believes

> The risk that high delivery frequencies will inflate transportation costs is overcome by ordering many different parts from each supplier, or by arranging for one lorry to collect each period from several different suppliers in the same district.

Such an approach would also be an economical solution in an MRP-based factory, but MRP would not make it as obvious that savings of this type should be made.

5.5.2 STOCKS PUSHED ONTO SUPPLIER

It is sometimes said that companies using JIT simply move the inventory to their suppliers. In order to prevent the stocks simply being held at suppliers and accruing cost through a higher unit price, closer relationships must be developed with suppliers. This may mean helping the suppliers to start the journey towards JIT manufacture or simply defining clearly what the demands will be and what notice will be given for changes.

Box 5.2   A note about push and pull.

The terms 'push' and 'pull' are commonly used to draw a contrast between JIT and MRP. They are imprecise and unsatisfactory terms. This box will point out the confusion that they can cause. (If you already use these terms in your company or with your colleagues, and have developed a satisfactory common understanding of what each means, you should ignore this box.)

Kanban is often termed a pull system, because the signal to produce more items is generated at the end of the product's flow path and causes each workstation to pull items forward from the previous buffer location. It can be thought of as a system where the customers stand at the end of the line and pull the products out as they require them. Items are therefore only made in response to customer demands.

MRP is often described as a push system, because work is issued at the start of its lead time and is thus seen to be pushed forward until it emerges from the factory at the due time.

Kanban can be seen as a push system, since the factory decides to hold stocks of items which it hopes the market will take. It is therefore looking back at previous demand and predicting that there will be more demand, so items are being made to replenish stocks in advance of customer demand, or to a kind of forecast. The controlling decision to make the buffer stock which always exists in the *kanban* system is taken in advance of any customer order, so this can be thought of as a push system.

MRP can be seen as a pull system since production is usually in response to a customer order. Nothing need be made unless it is required to be used to fulfil a customer order. The customer can be thought of as standing at the end of the line pulling items out, since nothing will be made unless a customer orders it.

A more useful treatment than the above may be to consider whether manufacturing takes place in response to actual customer orders or in response to forecasts. When manufacturing to order, as long as lead times can be reduced sufficiently and set-up times do not encourage large batches, it may be possible to eliminate safety lead times and safety stock (including *kanban* buffers) and to produce in the quantities required by the customers. When manufacturing to forecasts, it is tempting to cushion the business against the inaccuracy of the forecasts by using economic order quantities, keeping safety stocks and using safety lead times to cover unpredictable lead times. It is part of the role of the production manager to remove the long lead times, long set-up times and uncertainty which call for these uncompetitive inventory cushions. There will then be less need to manufacture to forecast, and manufacturing to order should mean that all excess stocks can be eliminated.

The question of push and pull is considered in more depth by De Toni *et al.* (1988).

5.5.3 INCREASED PAPERWORK

Increased numbers of deliveries would tend to lead to an increased administrative load and therefore increased costs. This would reduce the benefits achieved through holding less stock. Peters (1987) gives the example of Rank Xerox, who use simplified paper systems in so far as copies of a single document are used for ordering parts, payment, handling in house etc. This saves considerable volumes of paperwork and makes JIT easier to run.

It is not critical that all items be procured on a JIT basis. It is often not practical to procure small, cheap, frequently used items (sometimes called C-items) on a JIT basis. For these items it may well make sense to stock items and use fixed reorder point logic.

Most of the benefits of inventory reduction and JIT delivery come from optimizing the supply of the high value or high value of use components. These items are a small subset of all purchased parts. This means that realizing the majority of the benefits of JIT procurement is possible with only a slight increase in administrative overhead.

5.5.4 INFLUENCE OVER SUPPLIERS

Some writers argue that it is not necessary to have influence over a supplier in order to move together to Just-In-Time because it is of mutual benefit to both parties. Many companies have heard about JIT and many are moving towards smaller production batches, which means that they may be more able to supply as required. Increasing competition makes all companies more keen to satisfy the requirements of their customers. Nevertheless, in any particular case, the move to JIT is easiest where a considerable amount of business is done with the supplier and the customer has significant 'clout'. An example of this is given by Schonberger (1982).

In early 1981, the Kawasaki, Lincoln plant was buying steel tubing from a distributor. The distributor sold for a number of far flung tubing manufacturers, not one of which had an iota of a commitment to be responsive to Kawasaki. Over the summer Kawasaki buyers, quality controllers, and design engineers worked with Brownie Manufacturing Co., a local company that had acquired used steel-tube-manufacturing equipment. Brownie gradually improved the quality of its output, and in the fall Kawasaki negotiated a contract to buy about 30 percent of Brownie's output. From Brownie's point of view, the contract is large enough that it must heed Kawasaki's preferences for small, frequent deliveries and related services.

SUMMARY

Just-In-Time is a philosophy of good business based on the smooth matching of production to demand and the elimination of all waste. The control system most commonly associated with the use of Just-In-Time techniques is *kanban*. Both Just-In-Time and *kanban* were developed at Toyota by Ohno. The *kanban* technique is almost exclusively associated with manufacturing mass production commodities such as cars and televisions, while variants are available which are intended for environments of greater variety. JIT relies on a smooth demand and the removal of quality problems which allow safety stocks and safety lead times to be reduced. It also requires the establishment of a culture of continuous improvement and shared responsibility.

QUESTIONS FOR DISCUSSION

1. Why is it important to manufacture items in small quantities?
2. Just-In-Time calls for items to be prepared and delivered only just before they are needed. Why is this better than ensuring products are delivered in plenty of time?
3. In the calculator example, what information would need to be recorded on each *kanban* card for the single-card system? Design a suitable card layout and show how it would be filled in for complete calculators and for upper case mouldings. How many of each card would be in circulation in the example?

4. In the calculator example (Fig. 5.1), why are larger buffers used for purchased items and for items ready to be sold than for the items at intermediate workstations?
5. If the dual-card system were to be used, what information would be needed on each card? Produce a design for the cards and draw examples as above. Explain why each piece of information is needed.
6. In the calculator example, would it be practical to control the purchase of screws with *kanban*? If not, why not? What alternative method may be better, and what advantage(s) would it have?
7. If a company was thinking about implementing a JIT approach, how would you advise them to start?

REFERENCES

Bullinger, H.J., Warnecke, H.J. and Lentes, H.P. (1986) Towards the factory of the future, *International Journal of Production Research*, **24**(4), 708.

Burbidge, J.L. (1986) Period batch control, *Proceedings of the National Conference on Production Research, NCPR'86*, Napier College, Edinburgh.

Burbidge, J.L. (1987) JIT for batch production using period batch control, *Proceedings of the 4th European Conference on Automated Manufacturing*, IFS (Conferences) Ltd, Kempston, Beds, pp. 163–74.

Crosby, P.B. (1979) *Quality is Free: The Art of Making Certain*, McGraw-Hill, New York.

Deming, W.E. (1982) *Out of the Crisis*, Cambridge University Press, Cambridge.

De Toni, A., Caputo, M. and Vinelli, A. (1988) Production management techniques: push–pull classification and application conditions, *International Journal of Operations and Production Management*, **8**(2), 35–51.

Hall, R.W. (1981) *Driving The Productivity Machine – Production Planning and Control in Japan*, American Production and Inventory Control Society.

Peters, T. (1987) *Thriving on Chaos*, A.A. Knopf, USA.

Ohno, T. (1988) *Toyota Production System – Beyond Large-Scale Production*, Productivity Press, Cambridge MA, USA.

Sandras, W.A. (1989) *Just-in-Time: Making it Happen – Unleashing the Power of Continuous Improvement*, Oliver Wight Ltd Publications, Essex Junction VT, USA.

Schonberger, R.J. (1982) *Japanese Manufacturing Techniques: Nine Hidden Lessons in Simplicity*, Free Press, New York.

Shingo, S. (1985) *A Revolution in Manufacturing – The SMED System*, Productivity Press, Cambridge MA, USA.

Shingo, S. (1989) *A Study of the Toyota Production System from an Industrial Engineering Viewpoint* (transl. Andrew Dillon), Productivity Press, Cambridge MA, USA.

Van Loon, F.H. (1990) Into the 1990s: managing for customer satisfaction and profit improvement, *Quality Matters*, June.

Voss, C.A. and Harrison, A. (1987) Strategies for implementing JIT, in *Just In Time Manufacture* (ed. C.A. Voss), IFS Publications Ltd, Bedford.

Goldratt's theory of constraints (TOC) 6

6.1 INTRODUCTION

An important, original and influential view of production management was presented by Eliyahu Goldratt in the classic book *The Goal* written with Jeff Cox (1984). Described once as 'the world's best MRP love story thriller' (Shucavage, 1995) it sold millions, partly because of its original thinking and partly due to its being written in the form of a very easy to read novel.

The simple message of *The Goal* is that the success of the manufacturing business comes from determining the aim of the business and then dealing with the constraints which make the goal more difficult to achieve. As discussed in Chapter 2, Goldratt describes The Goal:

> The Goal of a manufacturing organization is to make money, ...and everything else we do is a means to achieve the goal. (Alex Rogo in Coldratt's *The Goal* (1993).)

To make as much money as possible, the company must strive to increase throughput while minimizing operating expense and inventory. Since the latter two cannot be reduced endlessly (they approach zero) the main avenue for development is in maximizing throughput. Goldratt explains that the element which limits throughput is constraining the system.

Goldratt's perspective on managing the production system is to identify the resource or 'bottleneck' which limits throughput and then to lessen its effects by scheduling the rest of the system in a way which prevents excess inventory and cost. These ideas were presented before *The Goal* in a scientific paper (Goldratt, 1980) which described the scheduling approach. They were also incorporated into a commercial CAPM package known as *Optimized Production Technology (OPT)* which was followed by a further development known initially as *Disaster* and then as *The Goal System (TGS)*. The descriptions in this chapter are mainly based upon the latest version, TGS. On the theory side, Goldratt developed the ideas of bottlenecks into a wider more general approach to management and problem solving known as the *Theory of Constraints*

(TOC) (Goldratt, 1990a) and the thinking processes described in *It's Not Luck* (Goldratt, 1994). It should be cautioned, however, that Goldratt's books are not easy to dip into, as they present ideas which are later seen to be incomplete or even fallacious as the full picture unfolds. Goldratt also tends to use terms in his own way and define them as he goes, which could confuse the reader who wants to look something up rather than to take the full course.

The specific details of the proprietary CAPM software are beyond the scope of this book to describe in detail, and in any case, the detail is likely to become incorrect as the software is developed. What is of interest here is the uniqueness of the approach taken by Goldratt.

6.2 CONSTRAINTS AND BOTTLENECKS

Goldratt's view of the throughput of a manufacturing system is that since it principally exists with the aim of producing products, its throughput will naturally increase until it is limited by some constraint, just as the flow of liquid from a bottle is restricted by the size of the neck. In his earlier work, which is mainly concerned with the flow of material through machines, the term 'bottleneck' is used to describe a machine which limits production. (Since all the components of a production system, such as resources, schedules, capacities and orders, are interrelated in a complex manner, it is often difficult to ascribe true cause and effect in identifying a single absolutely true bottleneck.)

In his later work, Goldratt tends to use the term 'constraint' to express the same idea in situations where throughput is limited by something other than a machine. Here, 'bottleneck' will be used for a bottleneck machine, and 'constraint' for any kind of limitation to throughput, including poor policy decisions or the market situation.

It is important to understand that the constraint can be anywhere in the manufacturing system, not just at the output end. Consider the example of a hosepipe carrying water, such as to water a garden. If someone stands on the hose, the flow of water will be restricted. If you are interested in the volume of water coming through the pipe it makes no difference whether the restriction is nearer to one end or the other, or whether it is in the middle.

In a factory situation, where a product must pass through several operations, the flow of products through the whole system is limited by the flow capacity of the slowest operation.

If operations upstream of the constraint continue at their full speed, a queue of work will build up in front of the bottleneck. If operations downstream try to continue, they will be starved of work.

The manufacturing system has also been compared to a chain, which, proverbially, is only as strong as its weakest link. It makes no difference where in the chain the weakest link is located. If the weak link fails, the whole chain is ruined, and a disaster might result.

These examples help us to understand that:

- the manufacturing system must be considered as a whole
- the throughput of the whole system is limited by constraints
- constraints can be anywhere in the system
- production at operations upstream of the constraint resources may produce excess inventory, while production downstream will be starved.

Goldratt made use of these concepts in the development of the original OPT production control package and its successors. These attempt to locate bottlenecks, to limit their effects, and to regulate the production of the entire system so that excess work in progress is not created upstream and to ensure that work is not ordered downstream which cannot be produced.

6.2.1 THE RULES OF PRODUCTION CONTROL BASED UPON CONSTRAINTS

The following rules are adapted from an article about Goldratt's approach by his colleague Bob Fox (1982).

1. The utilization of a non-bottleneck resource is not determined by its own potential but by some other constraint in the system

Consider the situation of a simple production system where a bottleneck machine feeds a non-bottleneck. The throughput of the whole system is limited to the rate at which the bottleneck can supply material, because the non-bottleneck can only process what it receives (Fig. 6.1(a)).

The reverse case is where the non-bottleneck feeds the bottleneck. In this case, the non-bottleneck can produce more than the bottleneck can use, but this can only build up as work in progress inventory. The overall throughput is still limited by the bottleneck (Fig. 6.1(b)).

A third case is where both machines separately supply parts for assembly, not to each other. In this case, assembly is limited by the bottleneck, since it needs a part from the bottleneck for each product. The non-bottleneck can generate excess work in progress inventory, but the amount of useful output it can produce is the amount required by assembly to match the output of the bottleneck (Fig. 6.1(c)).

Even when there is a single operation which has excess capacity and is not a bottleneck, its useful utilization is limited not by its capacity but by the rate at which the product can be sold, which can be seen as a constraint.

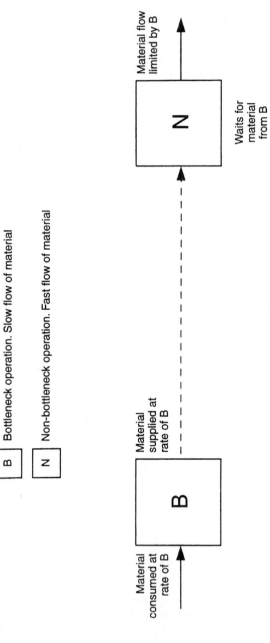

B — Bottleneck operation. Slow flow of material

N — Non-bottleneck operation. Fast flow of material

Material consumed at rate of B

B

Material supplied at rate of B

N

Material flow limited by B

Waits for material from B

Figure 6.1 (a) Material flows from bottleneck to non-bottleneck.

B — Bottleneck operation. Slow flow of material

N — Non-bottleneck operation. Fast flow of material

Material could be consumed at rate of N

N

Material could be supplied at rate of N

Inventory could build up!

Material consumed at rate of B

B

Material flow limited by B

Figure 6.1 (b) Material flows from non-bottleneck to bottleneck.

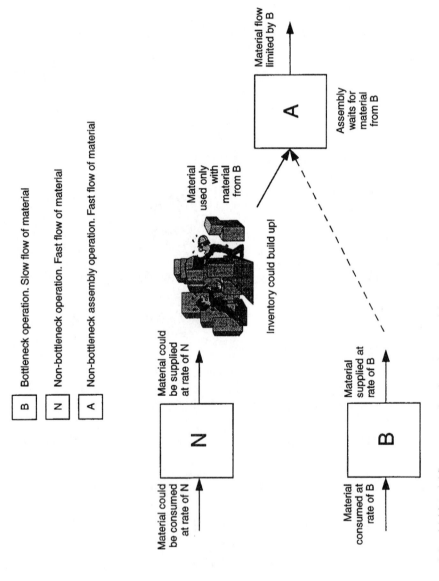

B — Bottleneck operation. Slow flow of material

N — Non-bottleneck operation. Fast flow of material

A — Non-bottleneck assembly operation. Fast flow of material

Material could be consumed at rate of N

N

Material could be supplied at rate of N

Material used only with material from B

Inventory could build up!

Material consumed at rate of B

B

Material supplied at rate of B

Assembly waits for material from B

A

Material flow limited by B

Figure 6.1   (c) Material flows to non-bottleneck assembly operation.

Thus constraints determine the amount of production that can be achieved by the entire system. The amount of work placed upon non-bottleneck machines must be determined by the requirements of the constraints. The only work centres that should consistently have 100% utilization are the bottlenecks, because these govern how much throughput can be achieved and what profits are made. It should be perfectly acceptable that non-bottleneck resources are not always busy, if the plant as a whole is running at the production level dictated by the constraints.

Inventory tends to collect where parts are waiting to be processed by the bottleneck, and where parts are waiting to be assembled with parts which come from the bottleneck. Machines downstream of the bottleneck will only receive items at the rate the bottleneck can process them. Rather than allowing excess inventory to build up, work should be scheduled in time with the bottleneck.

2. Activating a resource is not the same as utilizing it

To operate a machine or resource when the resulting output cannot get through a bottleneck, or when the output must wait for something else which is produced by a bottleneck, is to make waste in the form of excess inventory. The machine is active but not doing anything useful. Non-bottleneck machines should only work at the pace dictated by the rate of production of the bottleneck machines. Non-bottlenecks cannot pay for themselves by running all the time because the production of excess stock does not generate throughput. Rather, it wastes material, energy, labour, storage space etc. A non-bottleneck machine should always be available when needed to support the bottleneck, but it should not be expected to be busy all the time.

3. An hour lost on a bottleneck is an hour lost to the entire system

If there is a true bottleneck that is being utilized to its full potential, then an hour lost on this machine can never be recovered. The production which was lost during that time can never be caught up because the machine will never have time to make the items which would have been made in the lost hour, because it has no spare time. No matter how much stock was built up in inventory of related items, it cannot be used except with items from the bottleneck, so the whole factory might just as well have stopped for the hour. This shows how important it is to keep the bottleneck(s) running.

4. An hour saved on a non-bottleneck machine is a mirage

If a machine is not a bottleneck, it must have some idle time (or time which is wasted in making excess stocks). To save time at such a machine will increase the amount of idle time and will not increase throughput from the whole system, because throughput is limited by the bottleneck. Goldratt is very scathing of general cost-cutting and efficiency exercises which do not specifically target bottlenecks.

While throughput cannot be increased, it may be possible to decrease inventory by reducing batch sizes as long as the extra set-ups required do not then make the resource into a constraint. This has the extra advantage of reducing lead time.

5. The transfer batch need not always equal the process batch

Since set-ups use up the valuable capacity of bottleneck machines, it appears to be advantageous to group items into batches for processing together, so that set-ups are required less often. On the other hand, this causes an increase in the level of inventory and could cause a lumpy load for operations downstream of the bottleneck. This can be alleviated by using a certain batch size for the bottleneck process, but allowing the items to be passed on in smaller lots or in ones through the downstream machines. (This is known as the transfer batch.) This means that there will only be a build-up of inventory at the bottleneck, not everywhere in the system.

Those readers who wish to consider this in more depth should refer to Box 6.1.

Box 6.1   Constraints and economic batch quantity.

In the example quoted in *The Goal*, the factory would lose throughput worth $2735 if the NCX-10 machine stopped for an hour. If this is taken as the cost of performing a set-up, the EBQ would be the square root of $(2735D)/(CsV)$. If the value of the part is $10 and the other values are as given in the example in Chapter 3, the EBQ becomes 1403. For a machine with idle time, the cost of an hour's delay would be zero and the EBQ would be the square root of zero, as long as the idle time was not consumed entirely, in which case the cost of an hour's set-up would be the same as for the NCX-10.

These calculations are an effort to balance inventory cost against the cost of stopping production to perform a set-up. What they tell us is that the batch size on the bottleneck machine should be as large as possible, to minimize set-ups and maximize throughput, and on the non-bottleneck machine it should be as small as possible, to minimize inventory. If the figures are disregarded, that makes sense.

Since it is unlikely that there is a single customer order for almost two years' production, the 1403 batch is not likely to be feasible. We have disregarded the necessity to produce other parts on the machine from time to time – the part we have considered may be one of many which require the bottleneck resource, and it may only absorb a tiny fraction of the bottleneck's capacity. Imagine that the 1403 parts can be machined in a week. The hour spent making the last few of those parts, the ones which are required almost two years away, could have been used to make parts to put into products which could have been sold immediately for a throughput of $2735. How far ahead do you want to look, making things before they are needed, when you could make something else and sell it now? The answer to this question does not come from the EBQ. The need to make other things with the same machine is a good reason to make set-ups.

We have also disregarded the fact that the part might be used in an assembly with other parts, which might or might not have to wait as inventory, whose value is not taken into account by the EBQ calculation. Also, for a given throughput, large batches would increase the investment and lower the Return on Investment, which is overlooked in balancing inventory cost with set-up cost.

Most importantly, the value of the money invested in the part as inventory ignores the benefit of selling the completed product at a profit, although this relates to the whole product, not to any one of its parts. It is thus too simplistic to try to take one part or one machine in isolation, since profit can only be made from them when they are used with other parts and other machines. Both are component parts of bigger systems and useless on their own.

The only way out of this confusion is to regard the factory and the product as whole systems. The throughput earned by a product depends equally upon all its parts being in place, irrespective of their supposed individual value. The cost of operating a particular machine is a part of the cost of operating the factory. Operating performance can be calculated only by looking at the whole factory and the total throughput, inventory and operating expense. Operating decisions must be based upon the whole picture.

Question: The keen reader of *The Goal* may notice that Lou later corrects the cost figure for the NCX-10 to $2188. What difference does this make to the batch size to be used?

6. Process batch size should be variable, not fixed

MRP systems are often used with economic order quantities which become fixed for each item. These are calculated to minimize cost. In contrast, Goldratt advocates the most appropriate batch size to maximize throughput. The batch size should not be calculated on the basis of the part or the machine, but to allow the best throughput in any particular schedule, and any particular part may find itself with varying numbers of companions as it passes through various work centres.

7. Set the schedule by examining capacity and priority simultaneously

This recommendation is to some extent irrelevant nowadays, since different CAPM systems are able to use different algorithms to produce their schedules. The main issue is what to do with two different operations which fall on the same day. Early MRP systems using infinite capacity planning would leave the problem to be sorted out afterwards. Later MRP systems with finite capacity planning may give the time slot to the first job to be scheduled rather than to the one with the highest priority, and many of these schedulers would see each operation as independent of others and would not see the reduction in inventory and lead time which may arise from taking operations in different order.

Goldratt's scheduling routine takes into account all the constraints as it works backwards from the furthest point in the planning horizon. This is true backward scheduling, since it deals with everything required on that day before moving to the previous day.

In contrast, MRP sets schedules according to fixed lead times, and may allocate operations to many different days as it works back from the due date of the first item it happens to schedule. It must be determined afterwards whether an overload has been caused if infinite capacity planning is used and there is no clear guidance on what to do with it. If finite capacity planning is used it is possible that orders will be allocated time slots on the basis of which were dealt with first by the scheduling algorithm, with no consideration of the effect of their order upon the final delivery. It is possible for MRP to have allocated time to less urgent work before coming to something which is more urgent, but which may not easily fit into the schedule.

8. Murphy is not an unknown and his damage can be isolated and minimized

Goldratt uses the term 'Murphy' in a reference to the proverbial Murphy's Law – 'What can go wrong, will go wrong'. In order to create a reliable schedule, the effects of disturbances must be taken into account. Rather than allowing slack times throughout the system, which would increase inventory, Goldratt proposes the use of statistical methods to determine the likely scale and frequency of disruptions.

Where bottleneck machines are concerned, no allowance in slack time is possible, since it would have a negative effect on throughput. What is needed is to keep the bottleneck operating, which means protecting it from disruption. It is noted that disruptions tend to propagate through a chain of dependent operations, each being dependent on the lateness of the one before. It is essential to protect bottlenecks from disruption, which means that items should arrive ahead of time by the amount which is determined to be necessary protection. This is shown in Fig. 6.2. The release date for all the pre-bottleneck operations will be the same, indicating that after this time the operations should be carried out as soon as possible.

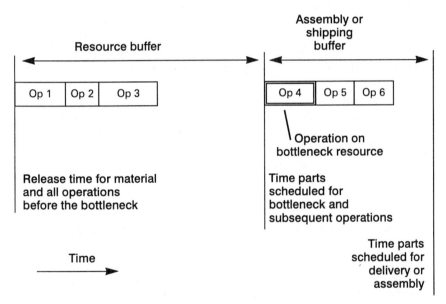

Figure 6.2 Buffers protecting bottleneck and assembly or delivery.

Goldratt's use of the term 'buffer' is close in meaning to the lead time used in MRP to time-phase orders before their due dates. However, whereas MRP assumes that the lead time is required to perform the operations on the item, Goldratt points out that the element of machining time it contains is usually a small percentage, the rest being made up of queue time and move time. Goldratt believes that if a suitable buffer is allowed, the item can be expedited if a disturbance uses up (for example) more than half the buffer time, so that the rest of the time can be used for the small amount of work which is required. Goldratt's buffer is therefore not directly dependent on the number of operations to be performed, and gives the time the material should be released for its first operation as the release date for all operations before the bottleneck.

A similar buffer time protects the shipping date from disturbance – this is known as the 'shipping buffer'. An 'assembly buffer' protects an assembly operation which uses parts from a bottleneck to ensure that the part from the bottleneck is not kept waiting for other parts to arrive.

Since the buffers are used for non-bottlenecks, which have idle time, there should be a good statistical chance that such expediting will work unless disturbances are extreme or very widespread, that is, when the factory is out of control. In the case of setting up a new scheduling system there is likely to be excess inventory and a backlog of orders which may mean that this reliability will take some time to achieve.

This logic is also based upon a certain amount of repetition in the factory, which can be used to provide a basis for the estimation of the correct buffers. If the factory produces a very wide range of completely different items using different production routes, this may present a problem. Fortunately, however, most factories are able to make estimates based on similar items, which should provide a reasonable guide. Nevertheless, it is still up to the production staff to ensure that the correct items appear in the correct sequence at the bottleneck.

9. Balance the flow, not the capacity

It has often been attempted to balance the capacity within a plant. This has usually been interpreted as keeping all the people and all the machines busy all the time. Plant managers attempt to keep expensive machines busy so that they 'pay for themselves' or because they are 'hungry'. These attempts lead to excess inventory and may ignore the true constraint. If the system is run at its maximum rate as determined by the constraint, all other resources will have spare capacity. If this capacity is used in the production of components which are sold as part of products whose flow is restricted by the bottleneck, the components can only be seen as wasteful excess inventory. Thus the production on all resources must be balanced to correspond to the flow through the bottleneck, not to their own capacity. In balancing the flow it must be acknowledged that many of the machines are not bottleneck resources and therefore will stand idle at times. The operators need not stand idle, since there is work to do in exploiting the bottleneck. However, it may be better for them to waste their time than to waste material as well.

6.3 GOLDRATT'S APPROACH TO DEALING WITH CONSTRAINTS

As a constraint, by definition, restricts the throughput of the manufacturing company, it is vital to understand each constraint and how its effects can be reduced before deciding the best way to manage production. When a constraint is identified, the throughput of the constraint – and therefore of the business – can be maximized, and efforts can be focused upon reducing the inventory and operating costs while maintaining that level of production. However, it must be cautioned that if too much emphasis is placed on the techniques of scheduling around constraints, which will be described in the next section, it may be possible to overlook the possibility of alleviating the effect of the constraint, and treating as fixed a constraint which can be removed by some creative thought or clever engineering.

There are therefore two sides to the management of production on the basis of constraints. The first is the identification of the constraint and the reduction as far as possible of its constraining effect. The second is the production and execution of manufacturing schedules which achieve maximum throughput from the constraint, and therefore the whole system, together with minimum inventory and operating cost.

Unfortunately, the two aspects are not entirely independent of one another. For example, if the mix of products being manufactured changes (which could happen because of changes in market demand) a production bottleneck at one machine may disappear and be replaced by another. A new schedule must be constructed to take into account the new constraint. Any change to the manufacturing system affecting its performance, such as a new machine, a new procedure, staffing arrangement, or information system could cause the constraint to change. It is therefore necessary to review both aspects constantly and to be aware of any changes.

The key distinguishing feature of Goldratt's approach is the identification of constraints, which govern and restrict production. Effort is spent on scheduling these resources to produce a schedule which will be reliable, while the other factors which do not restrict production are controlled by a system of lead times known as buffers and work lists to identify the sequence of work. Finer control is not required where excess capacity exists.

In dealing with constraints, Goldratt offers the following five steps, which are described fully in *The Haystack Syndrome* (Goldratt, 1990b). These steps should not be thought of as a strictly sequential methodology since iteration between steps may be required and – for example – constraints may be identified in various steps. The steps should be thought of as a continuous cycle of improvement which can be used over and over again, and which make continuous improvement a part of the process.

1. Identify the system's constraint(s)

While it is strictly true that there can only be one constraint at any instant – one weakest link – other constraints can appear when producing a schedule. If a schedule is produced to obtain the maximum throughput from the first constraint identified, another resource which has to process work for the same orders may turn out to be a constraint on certain days when its workload happens to be more than it can deal with, while over a longer period it may appear to have plenty of capacity.

It is equally possible that other constraints may emerge which were not initially identified. If one machine is overloaded by 50%, it may be possible to transfer some of the load to another machine, which may result in both of them being loaded to over 100%. There is thus a new constraint, and both must be considered in developing a production schedule. Besides being constraints on output, they are also constraints on the options available in scheduling. A procedure such as inspection could appear to be the constraint, but the true constraint could be the circumstances or policies which cause inspection to be necessary. If the constraint can easily be broken, the amount of throughput which can be achieved will be higher, increasing to the level at which a new constraint comes into play.

In the case where there is excess capacity in the manufacturing system, so that all orders can be met, the constraint may be the company's ability to sell. If that constraint could be broken, more products could be sold and more throughput could be earned.

2. Decide how to exploit the system's constraint(s)

Since constraints limit throughput, attention must be focused on achieving 100% of the possible throughput from resource constraints.

As a bottleneck machine is very heavily loaded, consideration must be given to ensuring that every moment is productive. This means ensuring that the work to be processed by the bottleneck work centre must always be available slightly before it is required, and never late (since lateness would lose time on the bottleneck, which is time lost for the whole system). It may be decided to hold a buffer in front of the bottleneck to protect it, as described earlier.

If the constraint is the market, it means that no order must ever be wasted, so 100% delivery on time should be achieved, while every step should be taken to obtain more orders.

3. Subordinate everything else to the above decision

The first two steps have focused upon identifying and exploiting the constraint. It is also important to look at how the other resources should be used. In controlling the factory this means using the other resources of the company in a way which helps to keep a bottleneck operating, thus ensuring that the bottleneck machine never has to wait for parts, and never breaks down, as far as possible. Thus the schedules generated should ensure that work which is destined for a bottleneck resource always appears slightly ahead of time. The schedule which is produced for the constraint is used as a basis to synchronize all other operations. Goldratt describes the constraint schedule as the 'drum', since the other operations are timed according to its beat. This is a reference to the scout analogy of bottlenecks, which Goldratt uses as an explanation in *The Goal* (Box 6.2). Thus all the non-bottlenecks are used in a subordinate role to maximize the output of the constraint.

Box 6.2 Goldratt's Boy Scout Walk analogy.

Goldratt uses the analogy of a group of boy scouts going on a walk to explain some of the ideas of throughput and inventory. Briefly retelling this analogy is the best way to understand the terms drum, buffer and rope as used by Goldratt. The full version is presented in *The Goal.*

The group of scouts are walking along a path. Each inch of the path is thought of as a product which has to be processed by each boy in turn as they walk in single file. This corresponds to a series of similar parts which must each be processed by the same set of operations in the same order. The aim is to process all the products, or for all the path to be walked by all the scouts. The different boys have different walking speeds, but as a group it is important that they all arrive at the same time.

When the slower ones are at the back, the faster ones tend to get away, and the line of boys stretches out. Inventory corresponds to the length of path which the first boy has walked over but which the last boy has not yet reached, that is, products which have been started but not finished. When the line gets longer, inventory increases. A simple solution is to put the slowest boy at the front and each faster one behind him, in order, until the fastest walker is at the back. The line tends to be short, and the path is processed at the speed of the slowest boy (Herbie). If there is any disruption along the way, such as when a boy stops for a moment to adjust his back pack, the boy who has stopped and the boys behind can easily catch up again because each can walk faster than the boy in front. Now the throughput of the whole system can be increased if ways can now be found to make Herbie go faster, such as relieving him of some of the weight in his back pack.

However, in manufacturing, it is not usually possible to make the first operation the slowest one and allow it to regulate all the others. The slowest operation could be anywhere. Imagine that Herbie is somewhere in the middle of the line of boy scouts. In this case, Herbie could beat a drum in time with his own pace, and all the others could walk in time with the drumbeat. Alternatively, a rope could be used to link Herbie to the boy at the front, so that the first boys would still be regulated by Herbie. Those behind him would be regulated by him anyway. Instead of the rope, it would be possible for Herbie to communicate with the front boy using signals to slow him down when he gets too far in front, or to tell him to speed up a little when Herbie starts to catch him. In manufacturing, the rope or signal represents a signal to release material at the pace of the bottleneck machine. The rope must be long enough to allow Herbie to keep walking if any boy in front has to halt for a moment, so that if a boy stops he will be far enough away so that Herbie will not catch up and have to stop. This is referred to by Goldratt as the *resource buffer* and is expressed as a length of time. It protects the constraint by ensuring that it will not have to stop. The length of the rope is determined by the likelihood of a boy having to stop, the length of time he is likely to stop, and the difference in speeds of the boys, which is the rate at which the buffer can be extended after a stop. The time the first boy has to start walking depends upon the time all the boys are to arrive, which is another way of saying that the customer order due date determines when manufacturing must start.

Other resources, such as expeditors, should also work to support the bottleneck. A repair team may be stationed near the bottleneck machine, to reduce the disruption caused by a breakdown. An hour's lost production on the bottleneck machine is an hour lost to the entire factory and cannot be recovered.

4. Elevate the system's constraint(s)

By 'elevate', Goldratt means 'lift the restriction'. Efforts should be concentrated on removing the effects of the constraint. If the constraint is a bottleneck machine, extra capacity may be found by routing some work to other machines or to subcontractors, or by buying extra resources.

Time spent on set-ups can perhaps be reduced, such as by using the SMED techniques of Shingo (1985), or by sequencing work to reduce the number and extent of set-ups. For example, a lathe may be used with a faceplate fixture for some jobs and with a jaw chuck for others. If the work is badly scheduled, time will be wasted in changing from chuck to faceplate and vice versa. Goldratt's scheduling tools deal with this time saving by evaluating the trade-off between the throughput created in the time saved and the excess inventory caused by bringing forward a later job so that two jobs can be processed together.

5. If, in the previous steps, a constraint has been broken, go back to step 1, but do not allow inertia to cause a system's constraint

If efforts to break the constraint have been successful, a new constraint will be restricting throughput, which will have increased to a new level. In order to continue to increase throughput, the new constraint must become the focus of attention. Production management must now use the new constraint as the bottleneck which must be exploited. Thus the improvement process must start again from step 1.

It is important to note that the whole system will operate in a different way now that it is restricted by a different constraint. This means that policies which have been developed in support of one constraint – to keep a particular resource operating, for example – may need to be altered when a new constraint emerges.

One feature of Goldratt's software is the very fast recalculation of the schedule, so that the removal of a constraint or the identification of a new constraint can quickly be taken into consideration.

It is possible for the system to be constrained by policies or ways of working which were appropriate in earlier circumstances but now become counter-productive. Inertia or the lack of change must not be allowed to invalidate the improvements made as the company develops.

This applies to any change in circumstances, not just those created by addressing the theory of constraints.

If the constraint is difficult to remove, for example, where a machine is running around the clock and the answer appears to be to invest in another one to provide extra capacity, this will take time. Until the new machine can be obtained, the constraint will continue and must be exploited.

6.4 MANUFACTURING CONTROL BASED ON CONSTRAINTS

Bottlenecks are important because the rate of throughput of the entire manufacturing system is constrained by the volume that can be processed at the bottleneck. Attempting to push more work through non-bottleneck machines simply causes increased inventory and queuing at bottleneck machines. This causes higher inventory costs and lower profits.

Every effort must be made to ensure that maximum throughput is achieved on the bottleneck machines. At the same time, inventory and operating expenses must be reduced throughout the system.

Attention to the control of inventory means that non-bottleneck machines will sometimes stand idle. The philosophy (like JIT) accepts that this is better than making things which are not needed yet because this in itself is wasteful. The whole production schedule is synchronized to the rate at which products flow through the slowest resource.

Attention to operating expenses means carefully controlling the use of overtime. Goldratt assumes that all staff are paid on an hourly or annual basis rather than on a piecework system, and that costs such as the heating or lighting bill are effectively constant. Overtime, however, is the element of operating expense which can be controlled.

Another key point of Goldratt's later work (*The Haystack Syndrome*) is the shift in thinking required in managers from the cost world to the throughput world. In former times, when labour represented a more significant part of manufacturing cost, it was possible to identify the cost of manufacturing different items and to associate profit with each product. Since labour is now not generally the principal element of manufacturing cost, systems which allocate the costs of running the business on the basis of labour hours create various anomalies which may result in inappropriate decisions.

Consider the example of a cheap, simple component used in the company's products. If the factory allocates overhead costs on the basis of labour hours, it may find that a supplier or subcontractor will offer to do the work cheaper. The cost-world decision would be to cease production of the item and to buy it instead. In the throughput world, however, the item would be looked at in relation to the constraints on throughput in the factory. If it does not take up time on a constraint resource, it does not limit the throughput of any other item, or of the factory as a whole. This means that it is occupying the spare time of resources which are not fully loaded. The only cost in making the item is the cost of the material. Can the alternative supplier supply for less than the material cost?

Goldratt's view that manufacturing systems exist to make money and that making money comes from maximizing throughput leads to policies and decisions which maximize throughput by identifying and dealing with constraints on throughput. The cost-world approach of putting cost considerations first leads logically to no business at all, which is the zero cost position.

It is not possible here to describe fully the detailed operation of Goldratt's scheduling products. A general overview of Goldratt's techniques should serve to identify their main principles. Bottleneck based schedulers have now also been produced by other software companies.

OPT and its successors produce factory schedules which help in the management of bottlenecks. The first stage of the exploitation is to identify the bottlenecks (Goldratt's steps 1,2 and 5) and the second is to produce schedules which manage production according to the flow through the bottlenecks (steps 3 and 4).

6.4.1 IDENTIFYING THE BOTTLENECK OR CONSTRAINT

A network model showing how different products use the factory's resources is created. An example is shown schematically in Fig. 6.3. The network shows each operation required on each part, the machine or other resource required, and the structure of parts making up the whole product, or various products. It thus combines the parts list with the routing information. The same file is used to store the inventory status of each item at each stage of its operation. This simplifies the data structure and allows faster processing times when compared to MRP which has to read the various data entities from different files (Chapter 4).

A calendar file is also used to identify working days and shift lengths which control the availability of each resource.

The schedule of customer orders is used by the software to calculate the load on each resource. This produces a profile of the load for each machine over a period of time, and the total load. Any resource having a load higher than its capacity is likely to be a bottleneck. Goldratt suggests that the orders which call for the highest loads on the supposed bottleneck should be checked, in case a high load has been generated by a typing error or some other mistake. The whole factory schedule will depend upon the correct bottleneck being identified.

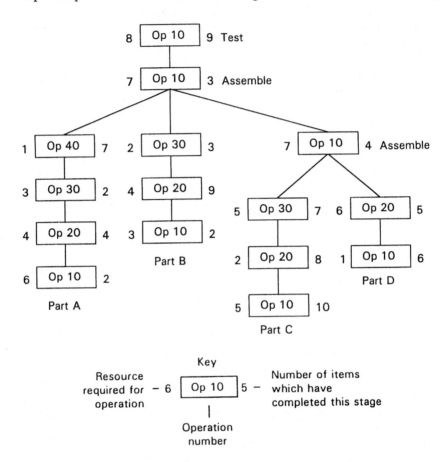

Figure 6.3 Example network.

At this point, it may be possible to re-route some of the work from the bottleneck onto other resources, thus balancing the flow and creating extra capacity, which may mean that some other resource is now the bottleneck. This is sometimes known as 'rolling the constraint'. It may also be possible to free some extra capacity by changing the sequence of work through the work centre to save time spent on set-ups.

Ways of taking work from bottleneck machines can also be considered. For example Goldratt and Cox (*The Goal*) describe a system where stress relieving is a bottleneck. (Stress relieving involves holding a component in an oven at a certain temperature for a period of time.) In order to increase the throughput of the entire plant it was decided to perform some cutting at a lower feed rate, which meant that stress relieving was no longer needed. The throughput of the system was increased even though longer was spent cutting. It did not matter because the load on the machines did not increase enough for them to become a bottleneck. An hour saved machining would be a mirage, as production would still be limited by the bottleneck.

The new load profile can be calculated, to see if the bottleneck is still where it was thought to be. Several iterations are possible at this stage. Eventually a bottleneck which cannot easily have work removed from it is identified, and the production route changes which have been decided upon are recorded.

6.4.2   EXPLOITING THE CONSTRAINT – SCHEDULING THE DRUM

The drum is the schedule for the bottleneck, which will later be used to synchronize operations on all the other resources. Scheduling is done by starting at the furthest point in the planning horizon and identifying the latest product delivery date. The required buffer to protect this date is subtracted (thus identifying the period in which all operations after the bottleneck operation must be performed). This gives the point at which the bottleneck operation must be finished and, by knowing the set-up and run times for the operation, the point at which it must start. (Since the throughput of the bottleneck is critical to the throughput of the factory, it is worthwhile putting some effort into ensuring that these figures at least are reasonably accurate.) This specifies a block of work on the bottleneck machine, with start and end points. The delivery date of the next-latest order is then considered in the same way, generating another block of work on the bottleneck machine.

If an item is required for assembly after processing on a bottleneck, the assembly buffer is allowed for in scheduling the drum, instead of the shipping buffer.

Each block of work is placed on a time line. Where a block overlaps one which has already been placed, it is shown graphically as if it was stacked on the other. Stacks of blocks may appear, some standing precariously upon the ones beneath. The screen which shows this in The Goal System software is called 'The Ruins'.

Next the stacked blocks must be dealt with. If the bottleneck resource is a single machine, it can only do one job at a time, so only one row of blocks is permissible. If the bottleneck consists of two similar machines, two rows of blocks will be allowed, and so on. Starting from the right, the later end of the time-scale, the blocks which are in an illegal row are pushed to the left until they drop into gaps. As they are pushed along, the sequence of blocks relative to each other is not changed, that is, a later block is not allowed to be placed before an earlier one. This means that blocks on the bottom row(s) will be pushed leftwards until the right-most block above drops into the gap on the right.

In the example shown in Fig. 6.4, blocks D, E and C are pushed to the left until D slots in between C and B, then blocks E and C are pushed to the left until E slots in between C and D.

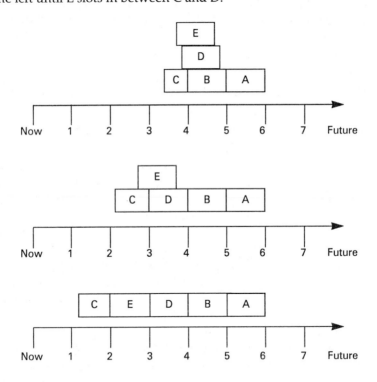

Figure 6.4 The Ruins.

When a bottleneck resource performs more than one operation on an item, a time interval must be left for these operations when scheduling the drum. This is represented in Goldratt's software by imaginary 'rods' of the required length, which are attached to the ends of the blocks to maintain the required separation between them.

All the operations now have a time allocation which is at or earlier than that which was originally required. However, since bottleneck resources are very busy, this leftward pushing is likely to result in some blocks being allocated times which are in the past. This is always a problem with backward scheduling. This can be dealt with by starting at the left and pushing all the blocks to the right until the first is scheduled to begin immediately, but no earlier. This has the effect of closing gaps between blocks, and may also result in blocks being pushed to the right of where they were originally placed. This is important to note, as it means these operations are now planned to be late. It may be possible to deal with these, as it may only mean that the buffer time for downstream operations is reduced. Since this is mainly an allowance for disturbance, careful allocation of priority to these items may ensure they are produced on time.

It may also be possible to shrink the schedule by:

- joining batches together to save set-up time (at the cost of increased inventory and the penalty of bringing one block forward and delaying all the others which are now after it)
- allowing overtime on the bottleneck machine (at the cost of operating expenses) if it is not already working around the clock
- using subcontract (at the cost of operating expenses).

Goldratt's software provides assistance with these decisions. If the lateness uses up all the buffer, and these measures are not available, the items cannot be delivered on time, but the customer can at least be informed. This should concern the items furthest away into the future, so that the customer is given the maximum warning of the difficulty.

The results of scheduling the drum are not evaluated until it can be confirmed that the non-bottleneck resources are able to operate in time with the schedule, since it may be that the schedule calls for an impossible load on a non-bottleneck at some instant. This may mean that another constraint has been identified.

6.4.3 SUBORDINATING – SETTING RELEASE DATES FOR NON-BOTTLENECK OPERATIONS

All non-bottleneck operations are now scheduled to synchronize with the bottleneck. These operations need not be scheduled to a high degree of accuracy, since the non-bottleneck machines have idle time available. What is required is:

- a release date after which work is allowed to proceed, which will establish priority when two pieces of work are available at the same work centre
- a profile of the load on the non-bottleneck work centre, since the schedule could generate unacceptable temporary overloads. (A difference between OPT and the later products Disaster and TGS is that OPT uses infinite capacity planning for non-bottlenecks, thus not providing a means of dealing with overloads on resources which have not been identified as bottlenecks.)

The network is now split according to the location of the bottleneck. Those operations upstream of the bottleneck, where inventory would tend to be overproduced, are termed the 'non-critical' resources. Throughput is not restricted by these resources as long as they produce the items required by the bottleneck. Although excess inventory must not be produced, there is extra time available to prevent any problems from disrupting the bottleneck.

At these resources, a certain amount of extra capacity is required to enable a buffer to be built up in front of the bottleneck to protect against disruptions or unexpected orders. The amount of this buffer has to be determined by experience, as does the acceptable time required to build up the buffer if it is used up. If the buffer is large and must be built up quickly to cope with a large amount of perturbation, this can add significantly to the workload of the upstream machines to the extent that they can become constraints.

The release date for all upstream operations is the date on which the item is required at the bottleneck less the buffer time. Subsequent operations between the first operation and the bottleneck operation will have the same release date, which means that once the buffer time starts, all those operations are authorized to proceed as soon as the material is ready, the release date acting as an indication of priority where more than one job is available at the same work centre. Intermediate dates would be meaningless, as the material is already committed and they would not help progress the material towards the bottleneck any more effectively. The release date for the first operation identifies the point at which material must be available.

The bottleneck and operations downstream of it may be regarded as the 'critical' resources, because they now constrain throughput. Obviously the bottleneck is the main constraint, but the items it produces must also be processed by the downstream operations. If any downstream operation is delayed, or if any item is damaged, throughput will be lost because the item can only be replaced by another from the bottleneck. Thus a problem at any critical resource limits throughput. The downstream operations, which correspond to the operations above the constraint in the network, are termed the 'Red Lane' in Goldratt's later work.

The date for all operations after the bottleneck is the same as the date allocated to the bottleneck operation, which means the item can now be used as soon as possible to make the finished product for the customer, to achieve throughput. When establishing priorities for downstream operations, the release date indicates the correct priority.

The shipping buffer sets the date for operations at and after the bottleneck, at and after assembly and for all operations on products which have no bottleneck operation or assembly. This operates even when the bottleneck operation is scheduled after the date given by the shipping buffer, so that the date establishes priority in the red lane when time is short.

A check on the load created at non-bottleneck work centres should identify whether unmanageable peaks of load have been generated by the subordinating procedure. Goldratt offers some approaches to these problems in his description of the routine for subordination, which takes into account the drum, the product structure and the load generated on non-bottleneck resources.

6.4.4 REITERATION AND RELIABILITY OF THE SCHEDULE

Since constraints may appear as the schedule is developed, the cycle of identify – exploit – subordinate can be repeated until all emerging constraints have been dealt with. Goldratt's software is claimed to be fast enough to allow many alternative schedules to be tried. Solutions such as re-routing or overtime can be attempted until a satisfactory schedule is found. (This is also claimed by many of the newer MRP-type systems.) As problems appear during the scheduling process, the schedule which is developed can be thought of as reasonably reliable, as long as it is based upon realistic assumptions about the performance of the organization, such as the required length of buffers.

The schedule can be managed by using expeditors to ensure the priority of work at non-bottlenecks, so that the bottlenecks are kept running and the maximum throughput is achieved from the system. By focusing upon the constraints, the approach identifies those resources that are likely to restrict throughput and cause increased inventory, and deals with them. It subordinates the other resources to a secondary role so that disturbances do not affect throughput.

While the use of buffers for non-bottlenecks is reminiscent of MRP lead times, the important difference is that they are only used where it is relatively safe to use them, not everywhere, and the results of their inaccuracies are decoupled from the factory's performance, not directly affecting it.

SUMMARY

The theory of constraints directs attention to the identification, exploitation and improvement of the constraints which limit the business' achievement of its goal. This is a different approach from that of MRP systems, which treat all work centres as being equally important, which, Goldratt teaches us, they are not. MRP systems make the assumption that a high standard of data accuracy can be maintained for all operations, and that data is used by the computer to derive appropriate decisions. While in some situations this is possible, Goldratt helps to identify the data which are important and ensures that the decisions are made by the users in an informed manner which relates to the key performance measures of the manufacturing business.

QUESTIONS FOR DISCUSSION

1. Define the term 'constraint'.
2. Define the term 'bottleneck'.
3. How can the performance of one or two machines affect the throughput of the whole factory?
4. 'An expensive machine should be kept busy as much as possible so that it pays for itself'. Is this a good rule of thumb?
5. If the theory of constraints allows some machines to stand idle some of the time, how far does it agree with or differ from the JIT approach to waste?
6. A colleague was visiting a company which runs on the principles of the theory of constraints. In his office, the Managing Director was explaining the throughput accounting system, which allowed him to see how much money had been made in the previous hour. The phone rang. It was a fork-lift truck driver from Despatch ringing to tell the MD that his car was a constraint on throughput, because the delivery van could not get past. Off he went to move it. What does this tell you about the company?
7. Do all companies wish to increase their throughput?
8. How is Goldratt's finite capacity planning different from that used in MRP?

REFERENCES

Fox, R.E. (1982) OPT – An answer for America, part 2, *Inventories and Production*, Nov–Dec. (The other parts appeared in the same journal as follows: part 1, Jul–Aug 1982; part 3, Jan–Feb 1983; part 4, Mar–Apr 1983.)

Goldratt, E. (1980) Optimized production timetables: a revolutionary program for industry, *Proceedings of the 23rd APICS Conference*, pp. 172–6, Los Angeles, October 1980.

Goldratt, E. (1990a) *Theory of Constraints*, North River Press, New York.

Goldratt, E. (1990b) *The Haystack Syndrome*, North River Press, New York.

Goldratt, E. (1994) *It's Not Luck*, Gower, Aldershot.

Goldratt, E. and Cox, J. (1984) *The Goal*, Gower, Aldershot (revised 1986, 1989; revised and extended 1993).

Shingo, S. (1985) *A Revolution in Manufacturing – The SMED System*, Productivity Press, Cambridge MA, USA.

Shucavage, D. (1995) *Crazy about Constraints* World-Wide Web page (http://www.lm.com/~dshu/toc/cac.html).

Process organization, product organization and group technology

7

7.1 INTRODUCTION

A factory is a complex system. The production manager is confronted by a range of machines and people with capabilities and skills, and a range of materials which must be purchased and items which must be manufactured, which may each require many operations. Different customers have different requirements, and different products may require different skills. All this complexity is too much for one individual to handle, except in the smallest of companies. The natural solution is to share the problem by assigning responsibility for different matters to different individuals, so it is usual for the management of the factory to be shared between a number of supervisors who are coordinated by the production manager. This chapter is concerned with the basis on which the factory should be subdivided.

In the higher volume industries, such as automotive or domestic goods, it is common to use flow lines for production, each line being concerned with a particular product or a narrow range of products.

In the medium variety area of batch and jobbing manufacture, factories have traditionally been run on the basis of grouping together similar machines and skill groups, which has often proved to be difficult to manage.

Group technology (GT) applies the grouping by products to the higher variety industries where products are manufactured in batches or as individual jobs. Group technology is not a way of managing production, but a way of making production easier to manage.

In each alternative, the problem of managing the whole factory is split into several smaller problems, each concerned with a part of the whole problem. Where the manufacturing company uses its skills separately, or in no defined order, the type of process may be the only obvious way to split up the management problem. Where the different skills have to be incorporated into products and are often combined in similar ways according to the nature of common products, it may make more sense to split the larger system into smaller ones on the basis of the products themselves. In this situation, it is important to achieve a good coordinated flow of material through the factory, rather than to achieve the optimum performance of each area.

Many companies naturally split their manufacturing resources into separate factories which deal with different types of product, which is a way of reducing the management problem at a higher level. Many companies also use a functional structure to split their entire organizations into sections which deal with different business activities, which may include such functions as manufacturing, design, marketing and accounting. New thinking in the field of business process re-engineering adds another alternative rationale for dividing up the business, this one based upon the flow of material and information around the organization. This will be looked at in Chapter 10.

The way the production management problem is divided up is affected by the mode of production involved. In Chapter 1, five modes of production were presented (Table 1.1). These were mass production, flow line, batch, jobbing and project. The choice of mode of production is a strategic one which must be made according to the aims of the company and its competitive strengths and weaknesses, along with its choice of what product to offer.

At the extreme of mass production, the company is most likely to be manufacturing a commodity product, which is by its nature difficult to distinguish from its competitors' products. At this extreme, the plant which processes the product is likely to be a dedicated facility through which the product flows as chemical or physical changes take place. Such products tend to be sold principally upon the basis of cost, and the choice of production mode is likely to be restricted to a question of making improvements to the process. The same process may run more or less unchanged, making the same product for many years.

Flow lines are used to provide the predictability of process plants for the manufacture of discrete items. They are principally used to reduce costs by producing large numbers of each item with great efficiency over a long period. Each line is dedicated to a product, or a small range of products. Product routings tend to be fixed.

Moving to the other extreme, the project mode of manufacture is used for one-off projects, where multi-purpose equipment may be brought in, or sometimes even designed and manufactured, for the specific project. There is no economy of scale because the same project is rarely repeated. The company using this mode of production therefore competes by bringing its specialist skills to bear on the project, often using highly qualified professional staff. The company is likely to have no catalogue, since it cannot repeat its projects. It can only use past projects as a demonstration of the expertise it is able to supply. The company's competitiveness does not depend upon the cost of products but upon the skills of its employees.

The same is true in jobbing manufacture. While there is an element of cost competition in any business, the jobbing factory often competes upon the extent of the skills it has available to perform a specific range of operations. It may produce products of its own, but these are likely to be ones which exploit the special skills of the company. For example, a company which specializes in long-bed turning produces a range of leadscrews for machine tools. As a very broad generalization, jobbing manufacturers are often in the business of providing services to other manufacturing companies on a subcontract basis, providing skills or facilities not available in the customer company.

It is in the area of batch manufacture that the most interesting production problems arise. Batch manufacturers who do a large amount of subcontract work are similar to jobbing factories, selling their skills in particular operations, while those who mostly produce their own range of products compete on the features, quality and price of their products. If the factory is arranged as a job shop, it may be very difficult to manage the production of whole products with long production routes consisting of a large number of different operations. If products sell in any quantity, the effect of any delay along the route is to generate a high level of work in progress inventory. Job shops, on the other hand, are easy to manage where only one or a few operations are required on a workpiece, since the production problems are mainly to do with the availability of a particular skill or facility at a given instant.

Group technology addresses the problem of the batch manufacturing company that produces whole products which are sufficiently different in routing to make flow line production impractical.

One of the main advocates of group technology, John Burbidge (1991), traced the first use of the term 'group technology' to Professor S.P. Mitrovanof of Leningrad University in 1961. He researched the relationship between part shape and processing needs. He found that savings in set-ups could be achieved on lathes by sequencing the work so that parts of a similar shape were processed one after the other. Since then group technology and various related approaches such as 'cellular manufacturing' and 'group assembly' have been applied in many companies and factories.

The traditional factory may be described as process-focused and the group technology factory as product-focused.

The traditional way of arranging manufacturing facilities has been to group together machines and processes according to their type. In the early years of industrialization this allowed specialists with certain skills to retain a craftsman approach, the skills being combined in products by the careful routing of products from one trade group to another, each trade operating as a separate business. This system worked well while there were only a few different products in the system, with clear relationships between the process areas.

As factories began to be established which would encompass the entire series of operations on a product from raw material to finished goods, the trade groups largely remained separate within the factory. As product ranges grew, especially in the 1950s and 1960s, the number of routes taken between process areas multiplied, with the effect that process areas no longer saw clear relationships with each other. This led inevitably to the use of route cards, progress chasers, WIP tracking systems etc., and with the growth of computer power led to companies putting their faith in more and more powerful computers to sort out the morass. Unfortunately, the imposition of a complex computer system upon a complex problem increased complexity yet further. This was recognized most notably by John Burbidge, who led a crusade against process organization.

Burbidge's alternative is to plan the production facility around the flows of material. This is called product organization, and leads to the subdivision of factories into areas often known as 'cells' (Box 7.1) in which a group of workers complete particular sets of products or components, which are then passed to a customer area for the next stage of manufacture. Thus within the group there must be a range of different skills which allow various operations to be performed.

Box 7.1   Cells.

Group technology work areas, comprising resources and the team of people who use them, are often referred to as 'cells' and group technology is often referred to as 'cellular manufacturing'. John Burbidge never liked the negative connotation of working in a cell, which, he said, was something we keep prisoners in.

An important benefit is that lead times within cellular manufacturing units are generally much shorter than in conventional process organizations. This is because the cell team automatically tends to schedule work dynamically through the cell, responding to any day-to-day occurrences which can upset a schedule set remotely over a fixed period. Also, the lead times within the cells become more stable, since the cell is less complex. This allows scheduling of work between cells to be more accurate. Cells are also the ideal vehicle for group improvement activity, especially since the team members have to pass their work on to their team-mates, providing instant feedback and peer-group pressure to maintain quality. This has the great advantage that the group becomes responsible for a tangible product and can to some extent control the way the product is produced. The cell is also a vehicle for the setting of improvement targets and for communication, consultation and participation.

In the design of their manufacturing systems, each manufacturing company must consider the choice of either process focus, based on skill grouping as in project engineering or jobbing manufacture, or product focus based around the flow of material to form products as in mass or line production. The choice is conceptualized in Fig. 7.1. The application of product focus in batch and jobbing factories can be achieved by the application of group technology.

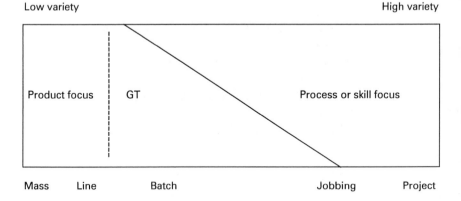

Figure 7.1 Product and process focus in relation to production variety.

7.2 PROCESS FOCUS

A process-focused machine shop is organized into a number of process areas. Similar machines, such as lathes, mills, drills or borers, are grouped together in each process area. The general nature of the process which is carried out is common to all the machines in the process area, and the machines may share common facilities and tooling.

For example, the grinding machines may be one process group. This has the advantage that the grinders themselves are together, as they can learn from each other and develop an in-depth knowledge of their key skill. This arrangement has obvious benefits for the training of new staff. In such a process-based group, all items which require grinding will pass through the grinding area, so the grinders can learn to apply their skills to any kind of grinding problem. There is also the advantage that all the grinding wheels can be kept together, and it may be possible to manage with only one of any type and size if they can be shared by the different machines.

The same advantages apply to process areas dedicated to other processes.

7.2.1 BENEFITS OF PROCESS ORGANIZATION

Development of key skills

The group of people who share the same type of skill can learn from and advise one another, so that the group as a whole becomes able to deal with any problem, or to develop a way to produce any feature which requires their skill.

Training and development

The skill- or process-based group is likely to encounter a range of different products which call for the application of the key skill, so the opportunity arises to teach the more difficult aspects to less experienced group members. The skills of the group members may be higher than the skills available through off-the-job training facilities.

Quality

Process specialists take a great deal of pride in the demonstration of their skills and will often make the effort to make their quality show, for example to produce excellent surface finish or to machine to spot-on dimensions.

Tooling

Groups which share similar processes will often require similar tools and equipment, such as machining fixtures, grinding wheels, lathe faceplates and chucks, cutting fluids and spare parts for machines. If these are located with the skill-based group they can be shared and the total number required will be lower.

Material

Some skill groups use common materials, which can be stored in a single location close to their place of use. This includes bar stock for turning and sheet metal for pressing.

Breakdown cover

If a skill group brings together all the equipment and machines which relate to that skill, such as all the lathes, the group provides alternatives which can be used if any machine is out of action for some reason.

Maintenance

The skill group can learn the best way to look after the kind of equipment which they use and, if necessary, maintenance teams can be assigned to the group, who become specialist in maintaining that kind of equipment.

Use of high and low skills

The grouping of machines of a similar type makes it easy for a highly skilled setter to perform all the setting and first-off machining for the whole group of machines, thus allowing lower skilled operators to be used to load and unload the parts from the machines as they are produced.

Legacy

There are clearly some advantages in continuing to use process organization if the existing resources are already set out that way, since it will avoid disruption and change. Many companies in the engineering industry have grown from a small beginning and have taken up the process form of organization. One may surmise that this may have arisen from a natural tendency to group similar items together, so new machines of a certain type are put with others of the same type, ignoring the less visible flow of products through the factory.

7.2.2 DRAWBACKS OF PROCESS ORGANIZATION

Since most parts which are manufactured require more than one process, they must be routed to and between the relevant process areas. This has two unfortunate effects on the material flow and the taking of responsibility.

The flow of material becomes 'jumbled' (Hayes and Wheelwright, 1984) since an item may move on from any process area to almost any other. This means that there may be no clear relationships between the process areas. While there are some common features of many routes – prismatic shaped items may often start with a milling operation; turning and heat treatment often precede grinding – these common routings may be insignificant if there is a wide range of items being manufactured. This means that the different areas tend to operate in isolation and to be unaware of the customer–supplier links between them, since these will be different for each item. The material flows form a complex network, as shown in Fig. 7.2.

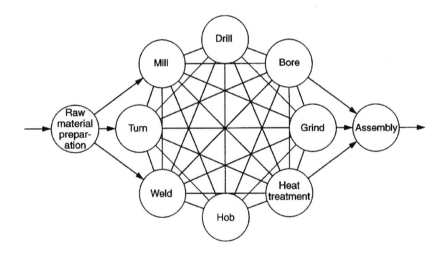

Figure 7.2 Material flow in a process-focused factory.

Since the material flows from area to area, it is difficult to say who is responsible for the manufacture of any particular item. If a part is late in arriving at assembly, it is not clear where the item may be held up. This leads factories which operate in this way to use sophisticated systems to monitor the progress of each item. These 'WIP tracking' systems often ask operators to register on a computer terminal each operation completed, often using bar codes or magnetic strips to help identify the items concerned. Nevertheless, since the flow is jumbled, parts may not be taken to their next operation as quickly as they should be, and the people responsible for the next process may not come looking for them before getting on with some other job which has arrived. Some factories use dedicated works movement staff to transport parts from area to area, which introduces another element of complexity. It is therefore common for there to be delays between operations. It also means that no operator or area foreman is responsible for the item; rather, they are responsible for machines within their areas. They are more likely to wish to keep their machines busy than to want to chase a particular item which has failed to arrive.

7.2.3 APPLICABILITY

It should be noted that these difficulties arise when the process-organized factory is asked to make complete products or complete parts. The process-organized factory is correctly configured to perform specialized operations as required, where items visit the factory for only one or two operations. If the factory is a subcontract jobbing business, where the key to the business's success is its ability to supply expertise in a specialized area, the process organization is ideal. The problem of material flow appears when a whole series of operations has to be performed. The company must ask itself whether its key competitive advantages come from being able to offer a range of specialized skills, or from being able to coordinate its production skills with those of design and marketing etc. to produce products which compete on their merits.

The design and construction of a process-focused machine shop is straightforward because of the distinct process areas (Burbidge, 1975). It is difficult to operate because of the very jumbled material flows and unclear responsibility centres as parts must be routed from one process area to the next. Complex centralized production control systems tend to be used to manage the activities to manufacture parts, and these are usually supported by distraught production managers and expeditors.

7.3 GROUP TECHNOLOGY AND PRODUCT FOCUS

The product-focused machine shop is arranged around groups of products or manufactured items and the groups of machines which are needed to make them. Parts which require more than one process to be manufactured enter a group of machines as raw material and leave as finished, or they may leave after completing a major stage of production. This requires the machines and support facilities within each group of machines to be capable of performing all the processes needed to produce the parts, so the machine group has to be devised to suit the part family. Looking at the whole factory, material flows are much simplified. The simplified material flows are the key to many improvements, as they determine the necessary complexity of the systems for functions such as scheduling, responsibility for quality and expediting (Burbidge *et al.*, 1991; Zelenovic and Tesic, 1988).

The complexity of a product-focused machine shop is reduced by the simplified material flows and clearly defined areas of responsibility. The design and construction of product-focused machine shops are more complicated because ways must be found to identify the groups. Also, common machines and facilities, such as pillar drills or deburring stations, may be required for components of many types, so in order to establish independent groups the machines must be duplicated. In many cases it will be necessary to change the routings of parts so that operations can be carried out on different machines.

7.3.1 FEATURES OF GROUP TECHNOLOGY

Burbidge (1979) identified group technology as a technique for establishing a product-focused organization and describes its introduction as:

> ...a change from an organization based mainly on process, to an organization based on completed products, components and major completed tasks.

The most important features of group technology are machine groups, part groups, ownership and team working.

7.3.2 MACHINE GROUPS OR CELLS, PART FAMILIES AND OWNERSHIP

A cell consists of the people, machines and facilities needed to provide the skills and processes required to take a range of parts completely through one or more major stages of production. In an engineering factory making metal products the major stages may be, for example, raw materials preparation (forging, casting or sawing), part production (machining or welding) and assembly.

Raw materials preparation includes all the activities required to take raw materials as they enter the factory and prepare them for use in the main factory. This may include such activities as the inspection and acceptance of delivered goods, the production of castings or forgings, and their inspection and non-destructive testing, the sawing of bar stock into billets, and guillotining sheet metal into blanks for pressing.

Part production includes all the operations required to make a finished component part from the casting, forging, billet or blank supplied by a cell in the previous stage. This stage includes all the operations needed on the part, including welding, heat treatment, plating and painting. It also includes any inspections which may be required to ensure that the parts passed on are completely finished.

Assembly involves all the operations required to produce a complete product, as required by the customer, using the parts made in the factory and any purchased items. Assembly includes subassembly, main assembly, test and packaging, so that the complete product is ready to be despatched to the customer or to the warehouse.

Within each major stage there may be one or many cells, but ideally each item should pass through only one cell in each stage. This results in a serial material flow between major stages and a parallel material flow within major stages. This is shown in Fig. 7.3.

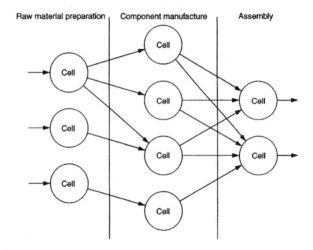

Figure 7.3 Material flow in a product-focused factory.

The parallel flow within stages is sometimes made difficult to achieve if there is a particular facility which is needed by more than one part family, such as electroplating. The effects of this sharing are to make the cells interdependent, which makes the factory more difficult to control. This will be dealt with later in this chapter.

As far as possible, each cell should be independent from all other cells. This means that no cell must share people, machines or facilities with any other cell (Burbidge, 1989).

The spectrum of parts that will have to be dealt with is divided into a number of ranges of parts. These ranges are part families. Each part family is produced completely from raw material (or taken completely through one or more major stages) by the team of people, machines and facilities of one cell. The parts in a part family share a number of similar attributes which require similar manufacturing activities. Such attributes could be shape within a size range, material or part type. The factory may be split into cells which are identified by different types of characteristics, so that some cells might produce items identified by their function, some by their volume, and some which are built around a particular type of operation, such as gear cutting. The important characteristic of the product-focused cell is that the items it produces are in some way complete, finished items. If a cell is organized around a gear cutting machine, so that all the gear and spline work can belong to the cell, the cell should contain all the machines and facilities to process all of its part family from start to finish, so that the cell still produces complete items. The amount of work performed by the special machine may then be quite small in proportion to the total amount of work done on the parts. Some examples of part families are given in Boxes 7.2 and 7.3.

Box 7.2   Some GT groups.

Company 1
Large cylindrical
   rollers and shafts
Small cylindrical
   rings, bearings, hubs, gears
Large prismatic
   frames, endplates, stands
Small prismatic
   blocks, brackets, spacers, arms
Spare parts
   listed replacement parts
Company 2 (Burbidge *et al.*, 1991)
Big gears
Small gears and shafts
Bar
Gearboxes
Drums
Plates
Turn

Box 7.3   Some interesting cells.

Magnesium-related assemblies
    fire risk, cell includes machining and subassembly
Lamination packs
    presses laminations and assembles them with other parts into
    subassemblies for motors etc.; includes coating, painting and
    machining
Goods inwards cell
    all receiving inspection and paperwork, including X-ray, and
    all material preparation for machining, including sawing and
    parts issue

Where it is not possible for the cell to control independently all of the activities needed to convey a part completely through one major stage, its ownership of the part may be expressed as a percentage, so that the overall design of the cells ensures that the ownership figures are as high as possible. This can be expressed as the proportion of the number of operations performed inside and outside the cell, or the value of the work in hours, or some other convenient measure.

7.3.3 TEAM WORKING

Each cell is manned by a team of people (Burbidge, 1989). The people may be the basis of the cell, particularly where the cell is a skill-based unit which provides a product based around a particular specialist skill, such as sheet metal working. The team of people may be self-managed to various degrees.

An opportunity to move away from work specialization and scientific management techniques can arise depending on how the teams are organized. With work specialization, tasks would be subdivided into independent plan, do and check sub-tasks in which different people specialized and were held responsible (Peters and Waterman, 1982; Peters, 1987). Scientific management defined 'one best way' for a specific task to be performed, matched people to 'worker' and 'managerial' functions, supervised, rewarded and punished these according to their performance and relied on indirect 'staff' groups to plan and control the task as advocated by Taylor (1947).

In contrast, team working can lead to the plan, do and check sub-tasks being turned into one task with the responsibility given to one team. Thus the works instruction as to how each item is to be made may be altered and improved by the cell team. In order to facilitate this development, the cell team may include people with production engineering and quality assurance skills. In most cases it appears that cells are managed best by a member of the team who is able to take part in the cell's work. In cells where the leader role is not formally assigned, a *de facto* leader often appears; equally, in cells where a leader with a management role has been appointed the leader has been seen to become an active member of the team.

7.3.4 PRINCIPLES OF GROUP TECHNOLOGY

According to Hyer and Wemmerlov (1984), group technology capitalizes on similarities in recurring tasks in three ways:

- 'by performing similar activities together, thereby avoiding wasteful time in changing from one unrelated activity to the next'. Since the parts in the part family share some attributes or features, they often require similar activities in manufacturing. If all the similar parts are made in the same cell, set-ups can often be saved by processing similar parts together, as long as the team of people in the cell have some freedom to determine the sequence of work for themselves. Depending upon the mix of parts in the family and the resources available, it may be possible to have machines or interchangeable fixtures permanently set for a particular feature, thus completely removing the set-up time.
- 'by standardizing closely related activities, thereby focusing only on distinct differences and avoiding unnecessary duplication of effort'. Each activity for any part family is performed by the same cell each time that a member of the part family is produced. This means that the cell team learn the best way to produce the parts, and can apply the best techniques for each part. Standardized activities can be developed as the activities for each part family are repeated in the same cell by the same team of people using the same machines and facilities. Over a period of time, the team of people will increase their 'know-how' associated with the part families manufactured by the cell.
- 'by efficiently storing and retrieving information relevant to recurring problems, thereby reducing the search time for the information and eliminating the need to solve the problem again'. The local control of activities in cells by teams helps the efficient storage and retrieval of information. As parts within a part family are all made and always made in the same cell then the need to repeatedly solve the same problems is removed.

An important point to note on the principles of group technology is that they capitalize on similarities in recurring tasks. The tasks themselves do not need to be identical. It is the similarities in any aspect between recurring tasks that can be exploited to gain savings. As an example, a number of different parts may be processed together on a drill, provided that they will each fit in the holding device (e.g. vice) and they each involve a drilled hole of the same diameter. The need to change the holding device and the drill bit is avoided and the number of set-ups incurred is reduced.

7.3.5 BENEFITS OF GROUP TECHNOLOGY

A great many benefits are claimed for group technology. These are generally from the point of view of a batch or jobbing manufacturer in the metal-cutting industry, making products out of machined parts. It should not be overlooked that the alternative, process focus, may be applied in other situations.

Reduced number and length of set-ups

The number of set-ups required, and hence the total time spent in preparation for activities, is reduced as parts which require similar activities are grouped together. The set-ups will also tend to be less involved and hence shorter as similarities between parts within a part family mean that the variance in their attributes is limited. This means that the preparation for activities involves only minor changes to accommodate the whole part family (Wemmerlov and Hyer, 1989).

Increased realizable capacity

Reductions in set-up times effectively increase the available capacity that may be realized from any given team of people, and set of machines and facilities (Schonberger, 1982).

Reduced numbers of fixtures and tooling

The team may develop specialized fixtures and tooling for the parts within a specific part family that will save time or serve for a range of parts rather than a single part. Reduced numbers of fixtures result in a further reduction in the number of set-ups incurred and require less space for storage (Wemmerlov and Hyer, 1989).

Reduced transportation and material handling

Each cell performs a number of activities on its machines and facilities that are geographically clustered. Distances involved for transportation and material handling are reduced. This saves time and reduces the opportunity for damage to and loss of parts in transportation (Burbidge, 1989; Schonberger, 1982).

Less expediting needed

The need for central expeditors to deal with a large number of individual people at separate machines or facilities is removed. Instead, to find and check any given part the expeditor only needs to check with the cell leader of the cell which owns the part. The cell leader will perform the role of expeditor at a local level, by knowing where the item is in the cell. In this way the problem of expediting is simplified and reduced (Burbidge, 1989).

Less production control needed

In a process-focused machine shop, every activity on a part must be scheduled and routed to the appropriate machine by the production control system. With group technology, only each visit to a cell by a part must be controlled by the production control system, since the cell can be given the responsibility of scheduling the sequence of activities which take place in the cell. Hence where more than one activity is performed on a part in a visit to a cell the total number of transactions that involve the production control system is reduced (Schonberger, 1982).

Less production planning needed

The purpose of production planning is to present effective methods to produce a given part, which has traditionally been the role of production engineers or planning engineers. In group technology, the team of people in the cell will become accustomed to effective methods to produce the part family which the cell produces. Once a certain amount of experience is gained by the team of people within the cell they will require less support from production planners (Burbidge, 1971).

Reduced inventory of WIP

A cell requires only one WIP queue at its input instead of one WIP queue at each of its machines and facilities, since work need not queue for each operation. Local control of the team of people and the machines and facilities allows WIP to be reduced between machines and facilities within the cell (Ranson, 1972). (This is only a clear advantage over a centralized production management system which attempts to schedule every operation on every machine. While this is very common in process-focused factories, it is possible to create a process-focused factory in which control is devolved to the process team, thus providing the same advantages of local control and reduction of inventory. This could be arranged along the lines of a cellular factory where each cell is based upon a particular kind of process, although some of the other advantages of group technology, principally the simplicity of material flow and the growth of knowledge of products, would be lost.)

Improved due date conformance

Reduced levels of inventory within the cells and the ability to move work immediately to the next operation mean that inter-operation queue times will be reduced, often from weeks to minutes. Hence the inter-operation queues have a less significant effect on the due date conformance of a part. The dominant factor that determines due date conformance becomes the length of the WIP queue for work waiting to enter the cell. The control of due date conformance is simplified, as only one WIP queue must be controlled, instead of all inter-operation queues. The clear responsibility each cell has for its parts also helps to remove a cause of lateness, when items are nobody's concern. Once a part is placed in a cell the team in that cell is responsible for meeting the delivery date (Ranson, 1972).

Reduced lead time

Reduced WIP levels within cells mean that parts pass quickly from one activity to the next until they are complete. Reductions in set-up times incurred allow smaller batch sizes. Smaller batch sizes can pass through the cell in a shorter time, so that the lead time is reduced, which in turn means that the level of WIP is reduced (Schonberger, 1982, 1986).

Reduced finished goods inventory needed

As the time taken to replenish finished goods is shortened, the level of finished goods inventory can be reduced. This reduces the risk of obsolescence, lessens the impact of engineering changes and lowers the investment in inventory (Wemmerlov and Hyer, 1989).

Reduced floor space requirements

Reduced WIP and finished goods inventory, together with reduced transportation and material handling, mean that less space is required to temporarily store and transport material between activities (Wemmerlov and Hyer, 1989).

Improved responsibility centres

In a process-focused machine shop, one part may be routed between several process areas and be processed by different people each responsible to a different supervisor. When a mistake is made or performance is poor, the responsibility for the part is divided between many groups of people. With group technology each cell is responsible for the production of a part. This should allow people within the cell to see the results of their work, because they manufacture a finished item, and it allows them to see and take responsibility for the quality and timeliness of the finished article. This should result in less 'passing of the buck' and better performance (Schonberger, 1986).

Improved quality (Box 7.4)

Quality is improved by a number of factors. The first is the responsibility on the team within a cell for the quality of the parts that they produce. A second factor is the reduction in variation (or the increase in similarity) of the attributes of the parts made in a cell, which removes some of the causes of error in manufacturing. Another factor is the reduced WIP levels and smaller batch sizes, which lead to the earlier discovery and correction of quality problems (Crosby, 1979; Schonberger, 1986).

Box 7.4   An avoidable problem?

A craftsman made a simple error and ground 123 pinions to the wrong diameter, scrapping a year's supply. If they had been in smaller batches, he would not have done as much damage, and if he had been used to seeing those parts every week he would have been familiar with the dimensions and might not have made the mistake.

Reduced independent inspection

Team responsibility for quality can be enhanced by self-inspection within the cell. This further reduces transportation and material handling, production planning and production control, thus further reducing the indirect labour costs. Lead times and inventory levels are reduced as a further WIP queue – waiting for inspection – is removed (Schonberger, 1982).

Self-maintenance

The ownership of part families by the cell can be extended to include ownership of the machines and facilities in the cell by the team of people. This is a logical progression as the team depend upon the machines and facilities of the cell to produce the parts for which the cell is responsible. Making the cell team responsible should allow them to do their own set-ups, tool design and maintenance. Usually, self-maintenance is limited to acts of little-and-often attention, such as lubrication, changing cutting fluid and changing lamp bulbs. The reasoning which is being applied in group technology is to devolve control to the cell team as far as possible, so that they can take full responsibility for their work. It is not suggested that major work can be done by the cell craftsmen, whose role is to produce a family of parts. However, when specialists are used for maintenance and repairs, it should be as a service to the cell craftsman, whose interest is to improve the machine. Where major overhauls are performed by maintenance specialists, it is suggested that team members can participate (Schonberger, 1986).

Improved job satisfaction

A team of people in a cell can be made fully responsible for the sub-tasks of plan, do and check for any task which was traditionally divided by work specialization and scientific management (Schonberger, 1986). This has been shown to have positive effects on the job satisfaction of teams of people in cells (Hyer and Wemmerlov, 1984; Wemmerlov and Hyer, 1989).

Increased people mobility and multi-skilling

If the people in a cell operate as a team, they tend to teach each other their skills and help each other. This has the effect of cell members learning a wide range of skills. If this is supported by training, members of the team can become fully skilled on a range of the machines or facilities. This will increase the flexibility of the cell (Schonberger, 1986).

Improved development of front-line people to management

Promotion of multi-skilled people from teams will result in management by people with a knowledge of the company's products and a wider understanding of more of the activities of the company than if single-skilled people were promoted (Burbidge, 1989).

Reduced indirect supervision

Self-management of teams reduces the need for indirect supervision. This can lead to the abolition of indirect supervision as the group supervises itself (Schonberger, 1982).

Reduced labour

As covered above, the features of group technology can reduce the need for transportation and material handling, expediting, production control, production planning, independent inspection, maintenance and supervision. Reductions in set-up times can result in more effective use of direct labour (Wemmerlov and Hyer, 1989).

Easier, more accurate cost estimates

The costing of products can be made easier and more accurate in two ways. Firstly, the complexity of costing is reduced as the cost of a product can be shown as a proportion of the operating cost of one individual cell instead of a proportion of a number of individual process areas (Burbidge, 1989). Secondly, a reduction in the proportion of indirect activities or overheads, such as transportation and material handling, production planning, production control, independent inspection and indirect supervision, will reduce the significance of overhead recovery rates. This can subsequently reduce indirect accounting requirements and further reduce waste.

7.3.6 DRAWBACKS OF PRODUCT ORGANIZATION

Difficult to plan

The creation of practicable groupings is difficult. This problem will be addressed in the next chapter.

Skill levels

Cell members tend to develop a broad range of skills, but may not have the depth of experience in any skill which could be required by an unusual job. They become expert in dealing with their product group, not in a particular operation. Whether this is a problem depends on the type of work in the factory, and whether there is a source of expertise to deal with difficult and unusual jobs.

Training

The cell may not offer a very wide variety of different work for people to gain experience.

Changes in product mix

Cell configurations may need to be changed if product mix changes. This may be seen as more of a problem in jobbing factories where machines are rarely moved. People who have experience in a production line environment are more likely to be used to reconfiguring resources from time to time.

Equipment and tooling

Some duplication may be required to allow each cell to have everything needed to completely finish a part family.

7.3.7 PREVIOUS APPLICATIONS OF GROUP TECHNOLOGY

Martin (1989): 'John Deere built their own group technology system in the late 1970s. They used mainly traditional machine tools as machine shops were reconfigured into cells. This proved to be very non-capital intensive. Currently at Deere, parts are completely manufactured in cells 80% of the time. Almost 100% of the time a part will be completely manufactured within a department via the utilisation of an adjacent cell'.

Hyer and Wemmerlov (1984): 'interest in group technology among US manufacturers took root in the mid 1970s. By the mid 1980s many large corporations such as John Deere, Caterpillar, Lockheed, General Electric, Black & Decker, and Cincinnati Milacron had implemented or were planning group technology programs'.

Wemmerlov and Hyer (1987): 'extensive use of cells was revealed as Japanese manufacturing became prominent. Their use was important in the achievement of Just-In-Time manufacturing. Group technology underwent a rejuvenation in Europe and the US as discussion took place in an attempt to learn and copy successful Japanese strategies. It had been demonstrated by the Japanese that cellular manufacturing can be a critical element in the rejuvenation of outdated and unproductive plants'.

SUMMARY

Process-focused units are the traditional way of organizing batch and jobbing factories. They provide high specialization for particular types of operation, but are difficult to control when whole products are to be made. They are especially prone to high levels of work-in-progress and long lead times.

Product-focused units are organized according to group technology techniques. Each group of machines is operated by a team who take ownership of a coherent group of manufactured items which have enough in common to form a family. Each group or team is usually given the responsibility to control the cell locally with the assistance of a cell leader.

The benefits of locally controlled product-focused units are wide reaching. They can largely simplify and reduce the support and monitoring functions that are normal with process-focused machine shops and also improve performance in terms of cost, lead time, due date conformance and quality.

In factories producing products, the advantages of using a product-based organization are enormous. The groups of machines and parts required can be determined by the use of group technology. This will be dealt with in the next chapter. The selection of an appropriate computer aided production management system for a product-based organization will be considered in Chapter 9.

QUESTIONS FOR DISCUSSION

1. What is a process-focused factory?
2. What is a product-focused factory?
3. How is the problem of controlling a large factory simplified?
4. What kind of factory would you recommend to be used by a company specializing in:
(a) manufacturing a narrow range of electric motors
(b) producing a wide range of gearboxes for all applications
(c) deep hole drilling in all kinds of metal
(d) subcontract manufacture of turned and machined parts used in other factories?
5. What are the advantages and disadvantages of allowing teams to control their own work?

REFERENCES

Burbidge, J.L. (1971) *Production Planning*, Heinemann, London.

Burbidge, J.L. (1975) *The Introduction of Group Technology*, Heinemann, London.

Burbidge, J.L. (1979) *Group Technology in the Engineering Industry*, Heinemann, London.

Burbidge, J.L. (1989) *Production Flow Analysis*, Oxford University Press, Oxford.

Burbidge, J.L. (1991) Group technology – definition, field and state of the art, *Proceedings of the International Conference on Production Research*, Hefei, China, August.

Burbidge, J.L., Partridge, J.T. and Aitchison, K. (1991) Planning group technology for Davy Morris using production flow analysis, *Production Planning and Control*, 2(1), Jan–Mar.

Crosby, P.B. (1979) *Quality is Free: The Art of Making Certain*, McGraw-Hill, New York.

Hayes, R.H. and Wheelwright, S.C. (1984) *Restoring our Competitive Edge: Competing through Manufacturing*, Wiley, New York.

Hyer, N.L. and Wemmerlov, U. (1984) Group technology and productivity, *Harvard Business Review*, Jul–Aug, 140–9.

Martin, J.M. (1989) Cells drive manufacturing strategy, *Manufacturing Engineer*, January, 49–54.

Peters, T. (1987) *Thriving on Chaos*, A.A. Knopf, USA.

Peters, T.J. and Waterman, R.H. Jr (1982) *In Search of Excellence: Lessons from America's Best-run Companies*, Harper & Row, New York.

Ranson, G.M. (1972) *Group Technology*, McGraw-Hill, New York.

Schonberger, R.J. (1982) *Japanese Manufacturing Techniques: Nine Hidden Lessons in Simplicity*, Free Press, New York.

Schonberger, R.J. (1986) *World Class Manufacturing: The Lessons of Simplicity Applied*, Free Press, New York.

Taylor, F.W. (1947) *Scientific Management*, Harper & Row, New York.

Wemmerlov, U. and Hyer, N.L. (1987) Research issues in cellular manufacturing, *International Journal of Production Research*, 25(3), 413–31.

Wemmerlov, U. and Hyer, N.L. (1989) Cellular manufacturing in US industry: a survey of users, *International Journal of Production Research*, 27(9), 1511–30.

Zelenovic, D.M. and Tesic, Z.M. (1988) Period batch control and group technology, *International Journal of Production Research*, 26(3), 539–52.

The identification of GT groups 8

Chapter 7 discussed the relative advantages of product- and process-focused factories. If product focus is to be adopted in a factory where an individual production line for each product is not feasible, a group technology approach may be used. The first difficulty with the use of product focus is the way to group the machines and part families.

Many authors have approached the problem of determining the best way to identify groups and part families. Many different computerized sorting techniques have been published, while some companies have had success in using their intuition and experience to group parts and machines.

The division of the range of manufactured parts and the range of facilities into groups has been called the Group Technology Configuration Problem (Askin and Chiu, 1990).

8.1 SCOPE OF GROUPING ANALYSIS

In any approach, the items under consideration must be determined. In a small company producing a limited range of products, it may be feasible to consider 2000 different parts being made by 20 or 30 machines. In companies where there is more variety, the number of items could be many times greater and it would be impossible to deal with them all, even with a computerized sorting algorithm. This is especially true of engineer-to-order companies, which continually generate new designs, and companies which continue to provide spare parts for old products. In these cases the grouping can be done by taking a sample of parts which are thought to be representative of the likely production over the foreseeable future period. Other existing items can then be allocated to the appropriate group next time they are called for, and new items can be designed with the groups in mind. Pareto analysis can be used to identify the sample, on the basis that a small number of products account for the bulk of the production volume. It is not unusual for the order book to be made up of a few items (say one-fifth of the total) which account for the bulk (say four-fifths) of the company's work measured in number of orders, work content or turnover. The one-fifth may be used as the basis for the initial division into groups, although it will depend on the individual case how much the results will be altered to allow for the other items which are expected to be required. An approach which was used in a capital goods manufacturing company was to base the design upon the company's entire production over an 18 month period, on the basis that this would give as good a representation of future loadings as could be produced in any other way, since all the products, though of the same general type, were engineered to order (Hallihan *et al.*, 1992). The extent to which information is available is also an important constraint.

It is also important to look at the whole of the manufacturing system concerned. If a factory is already split into departments, or perhaps two separate sites, each cannot be considered separately unless there is no item which is routed to operations in each department. 'It is impossible to divide a department which does not complete all the parts it makes into groups which do complete all the parts they make' (Burbidge *et al.*, 1991). Burbidge suggests that an analysis of the flows in a company and in each factory should first ensure that there is a simple unidirectional flow between all the departments concerned, which correspond to the 'major stages' described above. This may require parts to be re-routed. Company Flow Analysis (CFA) and Factory Flow Analysis (FFA) are described in Burbidge (1989). Groups within the major stages can now be found.

8.2 GROUPING TECHNIQUES

Four principal approaches to the problem were described by Burbidge (1971). These are division by eye, division by design classification, division by production classification and division by production flow analysis. To these can be added division by function.

8.2.1 DIVISION BY EYE

This simple approach involves the analysis of the range of parts by staff such as production engineers who have a detailed understanding of the production requirements of the parts. This may be done by simply printing drawings of all the items and grouping them by their similarities, which may be of shape, method or material. Piles of similar drawings are collected together, each item being added to the most appropriate pile. The staff concerned will develop some idea of the machines which will be associated with each group, and will gradually be able to decide which group a part belongs to. If a large number of small groups develops, these parts must be re-examined to determine whether they can be assigned to other groups. Groups of only a few items may not be viable if the workload they represent is only a fraction of the total workload of the machines which they require, since those machines may also be required to work on items in other groups. Routing information must also be available to allow this verification to be carried out.

8.2.2 DIVISION BY DESIGN CLASSIFICATION AND DIVISION BY PRODUCTION CLASSIFICATION

These two variations on the same theme both rely on coding systems used to describe components. Design classification systems are sometimes used to identify the features of an item from a design point of view so that designers can use or modify an existing part rather than having to design a new one when a new requirement develops. Production classification systems are sometimes used to describe the types of operation required to manufacture an item. In either system, each item is allocated a code number in which each digit stands for a particular attribute. For example, a digit may be used to indicate whether the shape of the part is prismatic (1), cylindrical (2) or a composite of prismatic and cylindrical shapes (3). The next digit may describe the material, the next the general size, etc. Examples of such codes are the Brisch code (Brisch, 1954) or the Opitz code (see Hyer and Wemmerlov, 1984). These codes can be used as the basis of an initial division of the range of items into groups. The groups of items must then be assigned machines, using a separate activity.

If such a classification scheme is not already in use, the effort and time required to classify parts makes this less attractive than to use a method based on part routings, which can virtually always be assumed to exist in some form. In any case, the groups which are proposed must be verified by reference to routing information.

8.2.3 DIVISION BY PRODUCTION FLOW ANALYSIS (PFA)

This approach looks at the flow of items between machines, based upon production routes. One type of PFA is proposed by Burbidge in his book of the same name (Burbidge, 1989). This technique in fact includes company flow analysis and factory flow analysis, which were mentioned above as essential considerations before any grouping is attempted. The actual groups are arrived at using group analysis, which is one of the many ways (Box 8.1) which have been proposed to deal with part routing information for each item.

Box 8.1 A pragmatic approach.

'...one of the leaders in organizing cells is a Rockwell plant, Telecommunications Division, in Richardson, Texas. In 1981, when it started planning cells, part numbers in its computer files were not coded for group technology sorting. Its method of grouping parts was to examine route sheets. Part numbers that followed the same route were a family. The next step was to clear out an area and move in the machines to make that family of parts.' (Schonberger, 1986)

Matrix sorting techniques

Various sorting techniques have been developed which use part routing information in different forms of matrices, as was proposed by Burbidge in his earlier work on PFA (1963), although his later approach is less mathematical and more manual (Burbidge *et al.*, 1991). In general, the idea is to rearrange the rows and columns of the matrix until groups of machines and/or parts appear. A sorted matrix is shown in Fig. 8.1. These techniques can be described according to the matrix axes, the matrix entries and the sorting algorithm.

Matrix axes may show either parts against machines, or machines against machines. In the part–machine matrix, shown in Fig. 8.1, sorting immediately shows both the machine groups and the part families. In this example, machines D, H and B will form a cell to produce parts N1, N9 and N7; machines E, C and G parts N4, N2 and N6; and machines I, A and F parts N8, N3 and N5. This is the simplest type of matrix to understand and use, but its limitation is the number of parts which must be considered. In a recent case the factory had over 100 000 live part numbers, which made such a matrix impossible to manipulate (Hallihan *et al.*, 1992).

In the machine–machine matrix, the result of the sorting is related groups of machines, which are generally much less numerous than parts. Although the matrix contains information drawn from the part routings, the part families must be identified by a subsequent stage to correspond to the machine groups.

Machine		Part number								
		N1	N9	N7	N4	N2	N6	N8	N3	N5
	D	1	1	1						
	H	1	1							
	B	1		1						
	E				1	1				
	C				1		1			
	G					1	1			
	I							1	1	
	A							1		1
	F								1	

Figure 8.1 A sorted part–machine matrix.

Matrix entries for part–machine matrices contain a logical figure which simply shows that the machine is used in making the part. Since the matrix is to be sorted by a computer, this means using the digit 1 or a logical 'true' depending on the software used.

Matrix entries for machine–machine matrices usually contain a figure known as a 'similarity coefficient'. A high value of the similarity coefficient indicates that the group of parts which use the machine on the x-axis is almost the same group as which uses the machine on the y-axis. If two machines process the same group of parts as each other, and no other parts, the similarity coefficient would be 1 or 100%, and the two machines should certainly be grouped together. If none of the parts processed by one machine is processed by the other, their similarity coefficient would be zero and they would not be expected to be found in the same group.

Similarity coefficients may be calculated in various different ways. For example, the simplest way of calculating the similarity coefficient between machine A and machine B would be to extract the routes for all the parts which visit machine A, that is, all the routes which contain machine A, and work out the proportion of those parts which also visit machine B. This calculation must be done once for each pair of machines, so if there are 20 machines there will be 190 calculations $((n^2-n)/2)$. Such a simple coefficient gives no regard to the different production volumes which will be required for each part, or the amount of work involved in each part, and therefore its importance. This can be a problem when the sorting is performed, because two machines which share a low similarity coefficient will fall into different groups. An important part which is made in high volumes and which causes a large amount of work should fall unquestionably into a cell which is equipped to make it, but if a very simple similarity coefficient is used, two of the machines required for the part could find themselves in different groups because the rest of the work they do makes them more similar to others than to each other. A more sophisticated technique would use the high work content and the high volume to raise the coefficient between the two machines, so that high volume, high work content parts will tend to draw together the machines along their routes. Such a sophisticated technique is presented by Gupta and Seifoddini (1990).

Another way of looking at volumes is to develop groups on the basis of volume, so that high volume items can be processed through a cell at high speed without being interrupted by lower volume items. Spares might be suitable for a separate cell on the basis of volume and their need for very short lead time. This removes some of the problems of volume by allowing the items with similar volumes to be considered using a less complicated coefficient.

Sorting algorithms are rather complex. The simplest type can be used with simple binary matrices. In this algorithm, each row can be read as a binary number, the columns of the matrix having different values, with the most significant at the left. The rows can then be sorted in order of the value of this number. Next, this is repeated column-wise (possibly by using a matrix transpose command), so that the columns are each read as a binary number, with the most significant digit being the one in the top row, and the highest value column being placed at the left. The procedure is repeated several times until the rows and columns cease to move. With luck, the matrix data forms into clusters around the leading diagonal of the matrix, which can then be boxed into groups. In the matrix shown in Fig. 8.1, both the row and column values are 448, 384, 320, 48, 40, 24, 6, 5 and 2.

A similar technique may be used with similarity coefficients, but in this case the row value cannot be read as a binary number, because the values are not only 1s and 0s. However, a value can be constructed by multiplying each coefficient by the value of the column it is in, and the totalling the value across the row. Those who are particularly interested in these sorting algorithms should consult McCormick *et al.* (1972), King (1980), King and Nakornchai (1982) and Chandrasekharan and Rajagopalan (1986a,b, 1987).

An alternative technique, which can be used with machine–machine matrices only, is to identify the two machines which have the highest similarity coefficient, and state that they will be grouped together. The two rows and two columns which relate to the two machines are then combined together into a single row and a single column, so that they are then seen as a single unit, which then has a similarity coefficient with respect to each other machine. The next highest similarity coefficient is then identified, and those two machines are combined together. If this works well, the individual machines gradually form themselves into larger and larger groups, until the last few amalgamations cause the groups themselves to merge into one big group. The results along the way should be recorded, and the different groupings produced can be assessed. If one iteration produces too many small groups, the next iteration can be used, in which two of them will merge. Groups of the required size can be found. One problem here is how the similarity coefficients should be recalculated each time machines are merged into groups. This technique is described by Gupta and Seifoddini (1990). If this technique works badly, each merger adds another machine to a single group, which simply grows until it includes everything. At each stage, there is one group and a remainder of individual machines.

Burbidge proposed a sorting method called 'nuclear synthesis' which generates groups manually using the routing information. Each machine is used as the key to generate a list of the parts which visit it and the other machines which are used to make all those parts, unless they have already been included in a previous module. In selecting the keys for the modules, the SICGE code is referred to (Box 8.2). The first key is the special machine with the lowest number of parts visiting it, then the next S-class machine, until they have all been dealt with, then the first I-class machine and so on. The groups are then produced by joining the modules to create groups. This has the effect of forming groups around special machines. The module contains similar information to a row or column of the machine–machine matrix containing similarity coefficients. A case study of the use of this approach is presented in Burbidge *et al.* (1991).

Box 8.2 Burbidge's SICGE code (1989).

S: **Special** Machines whose work cannot be transferred to other types of machines, and of which there is only one

I: **Intermediate** Machines whose work cannot be transferred to other types of machines, but there is more than one

C: **Common** Machines which are widely used and generally available in some numbers

G: **General** Machines or equipment used for a very wide range of parts and may well be required in more than one group, such as saws or painting

E: **Equipment** Items used to assist operations, such as benches or vices, which can easily be added to any group if required.

All these techniques depend upon the existence of a reasonable set of information about part routings. In many factories the routings are inconsistent, since they tend to develop over a long period when methods, equipment and planning engineers come and go. Whenever a new machine arrives, there is a great tendency to route all new jobs to it. This may be good practice if the machine is better than the older alternatives, but if older routings are not revised they become inconsistent (Box 8.3). The sorted matrix shown in Fig. 8.1 is deliberately over-simplified. In reality, there will always be operations which refuse to cluster, and which end up outside all possible boxes. In this case, either the part routings or the sorting technique must be re-examined.

Box 8.3 Inconsistent routings.

In a grouping exercise, three almost identical parts were found which had been planned during different eras of the development of the factory. One was routed to a mill, a drill and a jig borer, the next to a universal mill and the last to a CNC machining centre. If only the routes are considered, the three parts look different and might end up in different cells. The inconsistency appears in the number of operations, in the machines used and in the operation times.

8.2.4 DIVISION BY FUNCTION

This approach takes advantage of computerized information, which may be available as a relatively simple way of starting the grouping procedure, but it requires a good deal of thinking afterwards. The basis of the grouping is the function of the part, such as frame, shaft, roller, endcap, bearing, lever, cam etc.

The attraction of the technique is in the fact that where a production database of some form already exists it is possible to perform a preliminary grouping on the basis of the part's name in the database, without having to perform any complex matrix operations on routing information. It is also very appropriate to the principles of group technology, since it draws together items which can readily be thought of as in some way related as a family. This method is not subject to inconsistencies in the routing information, although it may show up inconsistencies, but by grouping on the basis of function it provides an incentive for items which fulfil a similar purpose to be manufactured in the same way.

However, it is a very imprecise technique, since the naming of parts is itself often very inconsistent. Similar parts may be described by different names, such as 'frame', 'bracket', 'support', and terms such as 'housing' or 'cap' may include a wide range of unrelated items. This analysis may be followed by some division by eye, and some consideration of the routings of the parts.

8.3 VERIFICATION OF GROUPS

It is inevitable that some problems will occur in the creation of groups. A common problem is for a particular machine to form a single group from two sets of parts which would be unrelated apart from the fact that they both use the machine in question. If this happens, it may be possible to allow the two groups to be split apart by routing one set of parts to an alternative machine. This may be a slower, older machine not generally liked by production engineers, but which may be able to produce the required output if limited to a product family which keeps set-up time low. The overall lead time, which is greatly shortened by using group technology, should be more important than the time which may be lost by using a slower machine.

Another common problem is the remainder group. After all the main machines and important parts have been allocated, it is not unusual to find a collection of unusual parts and machines which do not seem to fit into any group. Either both parts and machines must be forced into cells in which they do not really belong, or the parts must be subcontracted complete, allowing the machines to be allocated as spare capacity. An oddments or remainders cell is not recommended.

Problems always occur when facilities which can be neither duplicated nor split are required by more than one group. Typical examples are heat treatment, painting and plating. It should not be assumed that if parts require these facilities they should be allowed to go outside the cell. Every operation which is outside the control of the cell leader adds to the complexity of running the cell and reduces the chances of being able to deliver finished items on time. The first question to ask is about the nature of the operations which are required. In traditional process-focused factories, operations may be routed to any machine which seems appropriate to the production planning engineer of the moment, often without any thought for the material flow and lead time. It is not unusual for machined parts to visit the paint shop for a simple degreasing operation, involving dipping the component in a tank of solvent, simply because that was the only place the facility existed. However, with a little thought to the location of such facilities in respect of health and safety, it is not expensive to make these facilities available in the cell, a move which could save a week's lead time. Similarly, low-temperature stress-relieving operations, up to about 500 °C, can be moved from the central facility to the cell by equipping the cell with an oven. It seems pointless for work to travel to the paint shop to be marked with its part number in indelible ink, or to a central inspection area to queue up to be inspected. Deburring or fettling can also be moved to the cell, and if extraction equipment is to be made available for deburring it may as well be extended to allow some painting.

Subcontract operations are a problem on any routing, since they are another opportunity for the cell leader to lose control of work. Subcontract routings should be removed by either developing the capability to perform the operation in-house, if possible, or if not, by considering subcontracting the manufacture of the complete item.

A guideline is to aim for the configuration which provides the cells with the highest ownership figures.

All these considerations make the cell easier to manage by bringing all the operations on the parts in the family under the control of the cell team. Where it is absolutely necessary for parts to visit a shared resource, consideration should be given to designing a service cell with no part family of its own but with a contract agreement to return a part in a guaranteed length of time. At least if there is a guaranteed service, the other cells can allow the appropriate time and not lose control. If the service cell is given such a mission, it becomes an expert professional service with a clear role, especially if the trivial work is removed. Plating, painting and heat treatment may be dealt with in this way, as might the operations which are required during assembly, such as balancing or testing, which cannot be split between assembly cells.

The worst solutions are to make work visit another group which owns a vital piece of equipment, where it will have no priority, or to make the vital piece of equipment available to operators from any cell as and when they wish to use it. This results in clashes, disagreements and the rapid deterioration of the equipment.

Calculations of expected machine load must be performed to verify that the changes required to production routes do not produce overloads. This can occur for reasons of re-routing, but also when two or more identical machines are allocated to separate groups. The workload which they previously shared may now be allocated unevenly, since a certain amount will fall into each group.

Calculations of load should also look at likely production volume and mix changes to determine whether the groups are sufficiently robust.

The size of groups should be considered. Many of the advantages of group technology come from teamwork, and the group size should therefore be small enough to allow all the members to be recognized as individuals and to be able to communicate easily with one another. This suggests a maximum group size of around 20, although 12 to 15 would be preferable. A minimum group size should be set which allows the responsibility of a part family to be shared, especially at times of holiday and sickness, and which provides the group with a range of skills and experience to draw on. A minimum of six seems to work.

If a group is to operate over two or three shifts, the same maximum and minimum sizes should be used for the total of all the shifts. This means that the cell team will not be too large for the cell leader and all the members of the team to know each other. In shift working, it is useful if the timing of the shifts is contiguous, so that the work can be handed over direct from one person to another. With two shifts this also allows weekly meetings to be held with the full team. An overlap can be provided by extending the day shift a half hour at the end of the shift and the night shift half an hour at the start of the shift, one day per week.

8.4 WORK CELL DESIGN

Once each group of machines and facilities has been determined, it must be allocated an appropriate area of the shop floor, and within that area the most appropriate layout must be determined. Material flow should be considered at both the factory and the group level.

8.4.1 FACTORY LEVEL DESIGN

At the factory level, the groups may be arranged so that those working with raw materials will be placed closest to the goods inwards dock and the raw material stores, feeding work towards the assembly groups and the despatch area. Where the receiving and despatch docks are close together, the natural work flow pattern may resemble a 'U' shape. Nevertheless, the constraints of the building must be considered if the cost of moving machines is to be kept to a reasonable level. For example, if extraction or drainage are required in a cell, the design should try to make use of the existing facilities if they are available, possibly near the outside walls of the building. It is very expensive to install drains in the middle of a concrete-floored factory if no provision was made when the building was constructed.

Attention must also be paid to the existing location of each machine. Some of the larger CNC machines may be physically too large to move along the gangways, and it may not be possible to lift them over other machines to the most preferred location. Besides the supplies of electricity, compressed air etc., which must be made available in the new location, some of the larger machines must be placed upon a specially constructed concrete raft set in the floor to prevent any settling under the machine's weight. Considerable expense and lost production will be incurred if such a platform has to be constructed and such a machine realigned. Less weighty machines should still not be placed across the joins between sections of the floor. The design of the new factory should minimize these difficult moves, possibly designing the overall layout around difficult-to-move machines.

8.4.2 GROUP LEVEL DESIGN

At the group level, each group should be designed so that work proceeds in the most straightforward way through the cell. In some cases, the parts in the family may be so similar that the same production route is required for each item. In this case, the cell layout can be arranged in the form of a flow line, with conveyors or tables between the workstations to allow the smooth flow of items from one operation to the next. The flow line may be arranged as a straight line, or possibly in a U shape. The U shape brings the start close to the end, so people in all positions can be aware of the performance of the team as a whole. The U shape is also preferable from the point of view of communication between people. If the flow is very regular, people should be seated around the U-shaped work flow so that they can see each other, while if the work flow is less regular the work should flow around a central area within which the team can move from machine to machine.

Where the work flow is even less regular, that is, where the items being made take lots of different routes within the cell, it may be possible to place close together pairs of machines which are often used in consequent operations, so that as many as possible of the work moves are as short as possible. For example, a lathe may be placed close to a cylindrical grinder, since many items move directly from the lathe to the grinder. Inspection equipment may be placed near the operation which it inspects, so bore gauges may be placed on a bench near the borer, for example. A from–to chart may be used to help identify the most common flows between machines. This is a machine–machine matrix in which one axis represents 'from' and the other 'to'. In addition to the machines and facilities in the cell, there will be entries for the work flows into and out of the cell. The component routes are read, and a tick is placed in the appropriate box each time the work moves from one machine to another. The very over-simplified production routes which were used to create the sorted part–machine matrix in Fig. 8.1 are shown in Box 8.4 and the resulting from–to chart is shown in Fig. 8.2. From this it can be seen that there is a common flow of work entering the cell to machines A, D and E, from machine D to machine H and from machines G, H and I out of the cell. If real data was used, the numbers would have much more meaning than this simplified example.

Box 8.4   Simplified production routes.

Part	Route
N1	B, D, H
N2	E, G
N3	F, I
N4	E, C
N5	A
N6	C, G
N7	D, B
N8	A, I
N9	D, H

An alternative to the from–to chart is to use a pinboard with a plan of the factory, on which cotton is routed between pins to show the flow of work between machines. This is rather cumbersome for the design of the new cell, but it can be used with good effect to show the existing flows.

It is important that the team of people who work in the cell should be able to have a say in its design, since they will have to spend their time there. Some companies choose to equip each cell with a meeting table, where breaks can be taken, and provide the cell leader with a desk and whatever charts are required to show the progress of work and the schedules. The cell should have a clear entry and exit point for work flow, so that no job can find its way into the cell or out of it without the cell leader's knowledge. Space should also be allocated for tooling, personal lockers etc., and the cell should be arranged in such a way as to make it distinct from adjacent cells. Sundry equipment, such as benches, racks, jigs and fixtures, hoists, work bins and hand tools, should also be allocated.

From \ To		A	B	C	D	E	F	G	H	I	Out
Machine	A									1	1
	B				1						1
	C							1			1
	D		1						2		
	E			1				1			
	F									1	
	G										2
	H										2
	I										2
	In	2	1	1	2	2	1				

Figure 8.2 A simple from–to chart for nine components.

The team of people who will work in the cell must also be designed, as it is important to establish a team with the right mix of skills, and with people who are able to get along together. The cell should contain a mix of experienced and less-experienced staff, so that they can learn from one another.

It is important to take a holistic view in the design and implementation of cells. Where a piecemeal approach is used, for example to try one cell first and learn from the experience, it is possible for the first cell to take the best facilities and the best operators, leaving the last cells to be implemented with whatever and whoever remains. While implementation may proceed in stages, possibly starting with some experimental cells, the stages should lead to a final situation in which the needs of all the products have been dealt with and all the people and equipment have been allocated.

SUMMARY

There is a wide range of group technology techniques for the determination of groups. These suit all tastes from the pragmatic production engineer to the mathematical problem-solver. A great deal of time has been spent on deriving appropriate sorting and clustering methods, in an attempt to devise the optimum grouping based on the data available. However, it is inevitable that the set of cells proposed by the best of these algorithms will be subject to the scrutiny of all kinds of production staff who will identify problems which cannot be guessed from the routing information, and which may only arise when attempting to develop new routes for misfit operations. Adding to this the importance of producing meaningful cells which produce a credible set of items, it is clear that the use of intuition and experience should be very highly valued.

Computerized techniques can provide a starting point, but the results are often disappointing, and will always need to be adjusted. The aim of the grouping is to provide a smooth flow of work, to provide groups based upon the items the company sells, and to provide an environment where team working can improve quality and reduce lead times and inventory. If this can be provided by a manual approach which increases the involvement of the staff concerned, so much the better.

QUESTIONS FOR DISCUSSION

1. What range of parts should be considered in a grouping exercise?
2. What is 'division by eye'? Who should be involved in such an exercise?

3. What is 'design classification'? What is the result of a grouping exercise based upon component classification?
4. What is the result of sorting a machine–part matrix?
5. What is the result of sorting a machine–machine matrix?
6. When is a similarity coefficient used in the sorting of matrices?
7. Why might routing information be inconsistent?
8. What is 'division by function'? In what ways is this similar to and different from 'division by eye'?
9. Why should cells be independent of each other?
10. What factors influence the design of the factory layout as a whole?
11. What factors influence the design of the work groups or cells within the factory?

REFERENCES

Askin, R.G. and Chiu, K.S. (1990) A graph partitioning procedure for machine assignment and cell formulation in group technology, *International Journal of Production Research*, **28**(8), 1555–72.

Brisch, E.G. (1954) Maximum ex minimo, *Proceedings of the Institution of Production Engineers*, June.

Burbidge, J.L. (1963) *Production Engineer*, **42**, 742.

Burbidge, J.L. (1971) *Production Planning*, Heinemann, London.

Burbidge, J.L. (1989) *Production Flow Analysis*, Oxford University Press, Oxford.

Burbidge, J.L., Partridge, J.T. and Aitchison, K. (1991) Planning group technology for Davy Morris using production flow analysis, *Production Planning and Control*, **2**(1), Jan–Mar.

Chandrasekharan, M.P. and Rajagopalan, R. (1986a) An ideal seed non-hierarchical clustering algorithm for cellular manufacturing, *International Journal of Production Research*, **24**(2), 451–64.

Chandrasekharan, M.P. and Rajagopalan, R. (1986b) MODROC: an extension of rank order clustering for group technology, *International Journal of Production Research*, **24**(5), 1221–33.

Chandrasekharan, M.P. and Rajagopalan, R. (1987) ZODIAC – an algorithm for concurrent formation of part-families and machine-cells, *International Journal of Production Research*, **25**(6), 835–50.

Gupta, T. and Seifoddini, H. (1990) Production data based similarity coefficient for machine grouping decisions in the designing of a cellular manufacturing system, *International Journal of Production Research*, **28**(7), 1247–69.

Hallihan, A.J., de la Pascua, G.P., Childe, S.J. and Maull, R.S. (1992) A technique for the identification of group technology work centre groups and part families in a low volume, high variety capital goods manufacturer, *Proceedings of Sunderland Advanced Manufacturing Technology Conference*, Sunderland Polytechnic, 27–30 April, 1992.

Hyer, N.L. and Wemmerlov, U. (1984) Group technology and productivity, *Harvard Business Review*, Jul–Aug, 140–9.

King, J.R. (1980) Machine–component grouping in production flow analysis: an approach using a rank order clustering algorithm, *International Journal of Production Research*, **18**(2), 213–32.

King, J.R. and Nakornchai, V. (1982) Machine–component group formation in group technology: review and extension, *International Journal of Production Research*, **20**(2), 117–33.

McCormick, W.T., Schweitzer, P.J. and White, T.E. (1972) Problem decomposition and data reorganisation by a cluster technique, *Operations Research*, **20**, 993–1009.

Schonberger, R.J. (1986) *World Class Manufacturing: The Lessons of Simplicity Applied*, Free Press, New York.

Period batch control 9

9.1 INTRODUCTION

Period Batch Control or PBC is particularly associated with group technology and product-oriented or cellular factories. It is the control system proposed by John Burbidge for use with group technology. It follows the GT approach of splitting the manufacturing control problem into several smaller problems, each of which is allocated to an individual, and which is at a level of complexity which can be dealt with by an individual. PBC operates on the basis of providing a cell or group leader with all the resources required to manufacture a certain set of items, including an undisturbed period within which the group leader and his or her team have complete control over scheduling and the sequence of work. Inevitably, much of the content of this chapter is based upon the work of Burbidge.

At the simplest level, period batch control is a control system which assigns a volume of work to be manufactured in a fixed time period. It is usually used within a cellular environment, although it has been used in process-oriented machine shops.

Burbidge (1975) believes that 'Period Batch Control is fundamental to the successful introduction of Group Technology. It simplifies scheduling and capacity planning greatly and requires much less in the way of computing power to function effectively'.

The system is based on a fixed order cycle. A fixed volume of work is manufactured over a fixed time period. Work can only be included in the period's load if it is available at the start of the time period and it must all be completed by the end of the time period.

The items required over a time period are calculated by explosion of the bill of materials and the netting off of stock to calculate the finished goods required. This is very similar to the MRP process of calculating the gross requirements to fulfil the master production schedule as described in Chapter 4. However, because the lead times for all items are based on the standard periods this is a much simpler process than in MRP, where the calculation may be performed with a different lead time for each item. Whereas in MRP different lead times cause items to move in and out of phase with each other creating peaks and troughs of demand (Burbidge, 1970), the use of the standard period keeps items in phase with each other.

A schedule is developed by adding up the load required for each period until a suitable load for that time period is reached. This is a simple form of finite capacity planning. If the period becomes fully loaded, work may be shuffled around to provide a new schedule, or it may simply be added to a previous period which has capacity, a kind of backward scheduling. At the start of each time period the works orders are given to the group leader who decides in what order work should be processed within the period.

For John Burbidge, there was only one type of period batch control, which we will describe as 'standard' period batch control. This is the main type, which uses a single phase and a single cycle. Variations which allow multiple cycles and phases will be described later in this chapter.

9.2 'STANDARD' PERIOD BATCH CONTROL

From the point of view of the cell leader, the simplest form of PBC is as shown in Fig. 9.1. At the start of each period, the material for that period enters the cell, and it must all be finished by the end of the period.

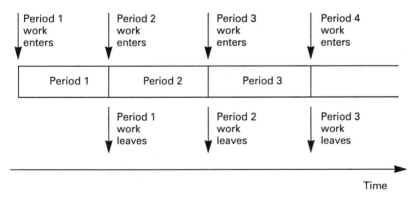

Figure 9.1 Cell view of 'standard' period batch control.

9.2.1 SETTING THE PRODUCTION TIMETABLE

From the point of view of the product, several cells may be visited, and parts and raw materials may have to be purchased. If the production flow in the factory is as suggested in Chapter 7 this may consist of the major stages of raw material preparation, component manufacture and product assembly. If, as Burbidge proposes (1989), each stage involves visiting only one cell, and the material can be purchased within a single period, possibly by a call-off arrangement (see Chapter 5), the products can be completed in a lead time of four periods. The four periods allow time for ordering, material preparation, component manufacture and assembly, as shown in Fig. 9.2. Each cycle of four periods begins with a scheduling meeting which determines which products will be manufactured for delivery four periods hence. Thus the customer orders which are dealt with at scheduling meeting 1, just before the start of period 1, will result in products which are completed by the end of period 4 and which can be delivered to the customer during period 5.

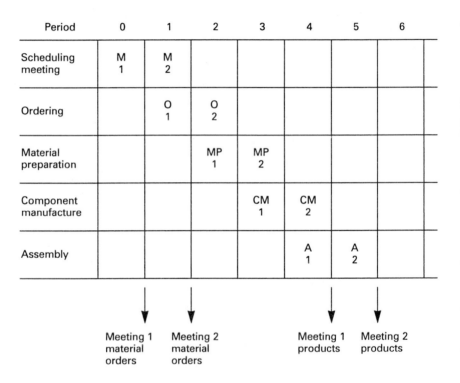

Figure 9.2 Production timetable for 'standard' period batch control.

This production timetable can be altered to suit the particular factory, but once set it should be followed for all parts and all cells. For example, it may be impossible to make arrangements for the material to be delivered in the same length of time allowed for the production periods, so two periods may have to be allowed, at least until a better arrangement can be put in place. If there is an extra major stage in production, such as a subassembly stage, or if the material preparation stage is not required, the timetable will be different.

9.2.2 SELECTING A SUITABLE PERIOD LENGTH

The length of the period is the most important variable in period batch control, so it should be chosen with care.

As in any production system, the amount of work in progress is broadly proportional to the lead time. As the lead time in PBC depends upon the length of the period, this means that a short period is desirable. It also allows a faster response to changing priorities. In the example described above, the scheduling meeting must plan four periods ahead. If production is to be to order, it is important that a lead time of four periods is short enough for the customer to accept.

On the other hand, production efficiencies can be gained by batching together similar items during the period to save setting time. If the period is longer, it is likely that more work will be available which can be processed at a single set-up, so there will be an overall saving in set-up time.

Another requirement of the period length is that it should be easy to work with. A period which is not a multiple of whole days or weeks would not be easy to use and would require a special calendar. In much of the literature on PBC (such as Burbidge (1988)) it is simply assumed that the period length will be one week. A week is a convenient length of time to work with, and it comes with the added advantage that it is followed by a weekend, which can be used if necessary to correct minor delays, thus providing additional robustness in the schedule.

In considering the customer's requirements, where products are made for industrial customers a lead time of four weeks is very likely to be acceptable. If shorter lead times are required, production is likely to be for stock, in which case a four-week forecasting period would be shorter than that in use in many companies.

The requirement to batch work together to save set-up time is only relevant if the set-up time is so long that it means that a short period cannot be used because the extra setting time would push the load on the cell beyond its available capacity, thus making it impossible for the work for the period to be completed in time. For this to be true it would mean that the setting time was a significant proportion of the cell's load. Rather than increasing the lead time and the work in progress to make long set-up times more economical, a better approach is to use the advantages of group technology and of having a dedicated cell team to investigate reducing the setting time, possibly applying the techniques of Shingo (1985).

If the setting up time on a machine is only an hour or so, one extra set-up only represents about a fortieth, or 2.5%, of the weekly load. In batch and jobbing factories it is very unlikely that any of the production figures would be known to such a degree of accuracy, so it is very difficult to argue that an extra set-up will overload the cell. Doubling the period length is unlikely to actually halve the average time spent in set-ups, whereas the application of the SMED system has been known to reduce set-up time to a few per cent of its original value.

Goldratt (see Chapter 6), points out that a time saving is a mirage unless it is at a bottleneck machine. The machines in the cell are not all likely to be fully loaded, so extra set-ups at the non-bottleneck machines are not a major loss. On the other hand, they allow a short production period with good service to the customer and low work in progress levels.

One other consideration is the total machining time of each item in the family. It is important, if not obvious, that the period should be long enough to allow each part in the family to be machined. The items with the longest total machining time must be given special consideration. This will be dealt with in the next section. If there are a few parts whose machining time is close to or longer than the total time available in the period, a multi-cycle system may be considered. This will be dealt with later.

Special arrangements must be made for parts which have to be sent out for subcontract operations. These should be in a small minority, and it is recommended that either the subcontract operations should be performed in-house or that the parts should be purchased complete.

9.2.3 RULES FOR WORK SEQUENCING WITHIN THE PERIOD

While PBC provides a smaller production control problem than that of attempting to schedule the whole factory, the cell leader will require some basic rules to allow a workable schedule to be worked out, at least when the cell is new.

The basic information which results from the scheduling meeting is a list of all the items which have to be made during the following period, including the quantities required. This is in addition to the information on batch cards which may accompany the material.

In order to be able to produce a schedule, the cell leader also needs a summary of the workload which these items represent on each of the machines and facilities in the cell. Burbidge (1988, etc.) calls this the 'period load summary'. This may be provided by a production information system which is centralized, and could be an adaptation of an existing MRP-type CAPM system. However, it is useful for the cell leader to be able to modify the information stored as methods change within the cell. It is essential for the operation of the PBC system that the cell leader can take charge of the cell, and this can only happen if the leader trusts the information which has to be used.

From the period load summary the cell leader can identify any machines which may be overloaded or close to their capacity. These are the first ones on which machine set-ups should be investigated, to determine whether enough time can be saved by processing together those parts which require similar set-ups or tooling. This is sometimes called 'follow-on scheduling'. The items can be listed in 'tooling families', and every attempt can be made to process these items in tooling families on the heavily loaded machines. This of course depends on their having completed previous operations on other machines, so the schedule must take this into account. Thus, the heavily loaded machines begin to set the schedule for the whole cell. The reader will notice the similarity between this consideration and Goldratt's 'drum' (see Chapter 6). If the time saved by tooling families is not expected to be enough to prevent overtime working, the cell leader should have a good idea at the start of the period of how much overtime will be needed each day, and appropriate arrangements can be made in advance.

The cell leader must also consider whether there are any items in the period whose total operation time is close to the amount of time available. Special attention must be paid to these items. Firstly, they must be started early. Next, they may be 'close scheduled'. This means taking the first item from the batch to the second operation as soon as it has completed its first operation, without waiting for the rest of the batch to complete the first operation. It should then proceed to the third and subsequent operations with no waiting time between the operations. The reader may notice the similarity between this and Goldratt's 'transfer batch' (see Chapter 6). This results in the creation of a temporary production line in which several machines will simultaneously operate on items from the same batch. If the route is a common one, the analysis of the flow of work used in designing the cell may have placed the machines in the correct sequence to make this easy. If the machines are automated, it may also be possible for the same operator to load and unload more than one of them.

The schedule must ensure that these 'critical parts' are started on as early as possible in the schedule, and that the machines which will be required for the second and later operations on them will be available when they are needed. Unless the critical items are very close in machining time to the total period length, it should be sufficient to operate by a rule that as soon as the first item of a batch of critical parts arrives on a machine, the next machine in the critical part's route should be set up for the critical part as soon as it finishes its current batch. This means the operators in the cell must operate as a team, and cooperate to manage the work flow.

Critical parts should be allowed to interrupt tooling families if the alternative is to fail to deliver a critical part. If a tooling family is interrupted, the effect is only to generate additional overtime, which is less serious. The prime concern is to deliver parts by the time they are needed. However, the cell team should be encouraged to operate flexibly, and no scheduling rule should be made so rigid that common sense cannot be used.

The cell leader should also consider the workload on the operators in the cell. Where there are more operators than machines, as is usually the case, the capacity of the cell will be determined by the amount of work the operators can be expected to complete. Since the cell is not expected to be inefficient, the overall load placed on the cell each week is likely to be close to the total capacity of the operators. This means that in order to achieve the schedule, operator waiting time must be minimized. This can be a problem at the start of the period, when all the machines which do first operations are busy and none of the others are. If there are some operators who are left idle, they should be allocated a second operation which can begin as soon as the first part completes its first operation. If possible, this should be an item where the second operation takes longer than the first, otherwise the second operator will have to wait for each part to finish its first operation. In practice, this does not appear to be a problem. (This is avoided in multi-phase PBC, described later.)

9.2.4 LOADING THE CELL

In calculating the load which can be given to the cell in each period, it is important to demand only a reasonable level of output. In particular, when the cell is newly established it is important to allow several periods in which the output is slightly lower than would be expected. This allows the team to become familiar with the part family and allows the cell leader to come to terms with the scheduling problem. If too high a load is placed on the cell when it is new, it can demoralize the team. However, once the cell is established, large improvements can be expected, especially when the cell is compared to a process-focused factory.

9.2.5 ADVANTAGES OF PBC

The principal advantage of PBC from the point of view of the factory as a whole is that it is a system which almost runs itself.

Cell leaders very quickly learn about their part family, their team and their equipment, and often output results have been obtained which have amazed manufacturing managers. Despite the lead times being fixed in PBC, they are frequently much lower than the lead times which result in MRP-type control systems (which are still the most common alternative). In turn, the level of work in progress is much lower and quality levels are often far higher.

Most importantly for the business, PBC brings predictability. As long as the cells are not allocated ridiculously high workloads ('let's see how much we can get out of them') the cell teams are able to manage the amount of work in a period and ensure that by teamwork, close scheduling and follow-on scheduling it can be achieved, sometimes even absorbing the disruption of machine breakdowns.

PBC works well because it is simple. However, the range of batch and jobbing factories includes some which have unusual problems. Variations of PBC have been designed to deal with some of these.

9.3 DIFFERENT TYPES OF PERIOD BATCH CONTROL

There are two characteristics that can be changed to tailor a PBC system to a particularly difficult environment. These are the time periods and the phasing of work.

9.3.1 THE PERIOD LENGTH AND MULTI-CYCLE PBC

In the standard form of period batch control, only the length of the period can be altered. All the work that enters a cell at the beginning of a period must be finished at the end of the period. This can be described as 'single-cycle period batch control', because all work goes through the cycle of production in the same length of time.

This length of time can be a problem in the situation where some of the parts in a family take far longer than others to manufacture. Multi-cycle period batch control allows multiple cycle lengths, so that some parts are completed in one period while others take two periods.

A multi-cycle period batch control system using two cycles is illustrated in Fig. 9.3.

For example, a turned parts cell may make pins and cylindrical parts. The longest throughput time for a cylindrical part may be one week and for a pin one day. The period length can be one day, with parts being allowed between one and five cycles for completion.

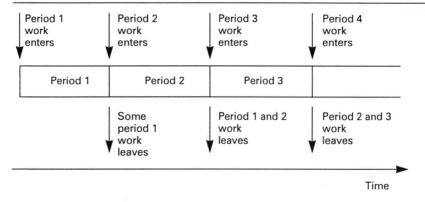

Figure 9.3 Cell view of two-cycle period batch control.

This system is much less easy to control from the cell leader's point of view because there are many parts with different requirements dates. The advantage is that short throughput time parts can be manufactured very quickly, while the benefits from batching work together are not available, and shop-floor-level control is reduced. This should only be considered if long cycle parts are unusual and it is impractical to provide a special cell for them.

An alternative, and much more attractive, method of achieving the same results would be to split the cell into two, one with a longer period and the other with a shorter period. This approach is only practical if there are enough machines and load for each of the cells. This reduces the complexity of the control system and allows each cell leader to control a homogeneous set of products. However, there is still disruption to the overall production timetable of the factory, since not all cells have the same periods, making the allocation of work to cells more complicated.

The eventual result of continuously shortening the time period is that eventually all parts have different cycle times. This means that it is very difficult to estimate the amount of load remaining in the cell at any instant, so that the amount of work that can be loaded to the cell at the start of a period will be difficult to calculate. By making the cell more complex, local control is removed and the advantages of the cell leader and a cooperative team are reduced. A PBC system where many parts have many different cycle times becomes rather like an MRP system and can be expected to suffer from the same problems. These are described in Chapter 4.

9.3.2 THE PHASE AND MULTI-PHASE PBC

In standard and multi-cycle PBC, the time periods during which work is scheduled run one after the other, each period ending at the start of the next (although in multi-cycle PBC some parts may run on into the next period).

Such systems may suffer the following drawbacks in certain applications:

- The minimum lead time in which a part can be manufactured is one period. Even when the cell is not busy the waiting time before work can enter the cell is up to one time period, and then there is the production period. This can be a drawback if orders for components have to be delivered very quickly, such as for spares. The lead time for parts is between one and two periods per major stage of manufacture.
- Where many items follow the same or similar routes there is a danger that at the start of the period the machines further down the route would be starved of work and at the end of a period the first machines would be starved of work. This occurs where the cell begins to operate like a flow line.

A solution is to implement a system which has time periods that overlap. This is known as 'multi-phase' single-cycle period batch control. This will allow work into the cell more frequently, while maintaining a volume of mixed work in the cell which will even out the machine loads and allow cell leaders to achieve benefits such as set-up reduction by follow-on scheduling. The simplest form of multi-phase system is dual-phase period batch control. Dual-phase period batch control is illustrated in Fig. 9.4. In a dual-phase system there are always two cycles of work in progress.

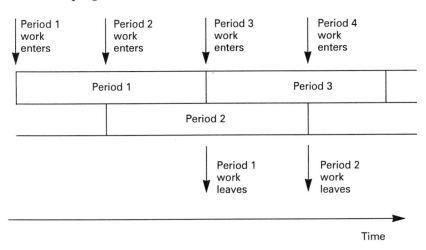

Figure 9.4 Cell view of dual-phase period batch control.

In the dual-phase system, there will always be two sets of work in the cell. This means that each period must be loaded with half the amount of work that would have been loaded under the standard PBC system. If the time period is one week, half a week's work would be inserted into a cell twice per week, one set of work arriving at the start of the odd-numbered period and one at the start of the even-numbered period. Each item in the set of work still has the same period, so each set of work is due for completion one period (week) after it arrives in the cell.

By providing a mix of work in the cell, the problem of the workload moving from the starting machines to the finishing machines is removed. There is always some work in the first part of its period and some nearing completion.

By overlapping the periods, the waiting time before a particular item can be given to the cell is reduced to half the length of the period. (If there were three phases, it would be one third, and so on.)

Since the two (or more) sets of work have different due dates, the cell leader must be provided with a simple way of determining which work is for which due date. A simple method was proposed by de la Pascua (1992), which used three separate coloured tags to identify work for two phases. At any instant there will be two colours in progress. Any work which shows the third colour is clearly late, since the third colour should be out of circulation. The operation of the colour system is illustrated in Fig. 9.5.

Since the second phase does not start until the middle of the first week, consider the second week.

At the start of week two there remains some red work, which is due for the end of period 2. The cell is loaded with a consignment of work equivalent to the capacity of the cell for one half of one week. A whole week is allowed for this work. It is given a blue period indicator, and is due for completion by the end of period 3.

Half-way through week two a further consignment of work equivalent to the capacity of the cell for one half of one week is loaded to the cell. This work is to be completed by the end of period 4, which is half-way through the next week. It is given a yellow period indicator. By this time the red work should have been completed. The cell now has only blue and yellow work, and any work still in the cell which has a red tag is late.

At the start of week three all work with a blue period indicator must be complete. The next consignment of work is loaded to the cell. This work is to be completed at the end of period 5, the end of week three. It is given a red period indicator.

Multi-phase systems are useful in cases where the throughput times of the components are high and lead time is critical. In these cases the lead time cannot be shortened by shortening the time period because of the long throughput times.

Period indicator	Week one		Week two		Week three	
Yellow	Period 1					
Red		Period 2				
Blue			Period 3			
Yellow				Period 4		
Red					Period 5	
Blue						
Work in cell	Blue (not shown) + yellow	Yellow + red	Red + blue	Blue + yellow	Yellow + red	Red + blue

Figure 9.5 Operation of the dual-phase cell control system.

Multi-phase systems are useful in three main instances:

1. In single-phase systems where lead time must be reduced and machining times are too long to allow further reduction in the length of the cycle.
2. Where a large volume of work is required in order to batch work together to prevent the effect of long set-up times.
3. Where most of the work goes through the cell in a similar route. A single-phase system would starve some work centres of work at the start and end of a time period.

These advantages must be weighed against the additional complexity which is involved. Where waiting time is a problem, an alternative may be to establish a particular cell for those items. Where set-up times are a problem, they should be reduced by the application of SMED techniques (Shingo, 1985). In discussion with the late John Burbidge, neither of us could think of a case in which the problem of work centres at the start and end of the production route being starved of work had arisen. Thus, the arguments in favour of this form of PBC make its useful range of application very narrow. Nevertheless, this system has been used successfully in the case described by de la Pascua *et al.* (1992), where an important requirement was to reduce the waiting time for urgent parts.

If a multi-phase system was combined with a multi-cycle system, the meaning of the periods would be lost completely, and in-cell control would be almost impossible. This would be practically the same as the MRP system.

9.4 LOCAL AND CENTRAL CONTROL

The control of the cell is passed to the people in the cell. This is good because conflicts of interest between different jobs are reduced – all the jobs will be completed in the period, so it is not necessary to determine which is the most urgent and to process it first. The cell leader can use the time available in the best way to ensure the required output.

A strict discipline is needed to ensure that works orders enter cells only at the start of a time period or wait until the next time period. If expeditors or sales staff are allowed to place additional demands on a cell during a period, the cell leader's control is reduced and chaos may result. The control within time periods is within the cells, but the control of the work entering the cell at the beginning of each time period is centralized.

PBC is usually associated with a cellular environment. The calibre of the people within the cells is critical to the success of the cells. The output which can be achieved and the continuous improvement which is possible depends upon the people in the cells. This is not true of systems such as kanban, MRP and OPT, where the computer system or the order in which cards arrive decides the order of work in an area.

SUMMARY

Period batch control provides an environment within which a team of people can take the best decisions to ensure the required level of output is achieved, even when product mix and machining requirements change. Splitting the scheduling task into cells and time periods reduces the complexity.

Burbidge (1988) suggests that '...the introduction of group technology (GT), coupled with period batch control (PBC), so simplifies the problem, that with minor analytical assistance the GT foremen can do their own operational scheduling, more reliably and cheaply than can be done by the computer'.

Local control is responsive to problems and can ensure output even when breakdowns and other problems occur. Knowledge regarding the strengths and weakness of the people and machines can be used in making informed scheduling decisions. By allowing total flexibility within a firm boundary, the system achieves both local control and predictability of output. This can be likened to the 'simultaneous loose–tight' properties of Peters and Waterman (1982).

'Standard' period batch control provides a viable control system for almost all batch and jobbing factories, as long as they are arranged along the lines of group technology.

Where there is a higher variance in lead time, and lead time is critical, a multi-cycle system allows some parts to be manufactured with different throughput times.

If the minimum throughput time is sought, a multi-phase system provides a shorter waiting time for work to enter the cell.

QUESTIONS FOR DISCUSSION

1. What is a period?
2. What is the 'production timetable'?
3. How does period batch control simplify the problem of scheduling in a factory?
4. What are the advantages of scheduling work on the shop floor?
5. What must the cell leader look for in planning the work for the period?
6. What three things must be taken into account if the cell is to achieve its required output?
7. What are the advantages of working with a schedule which looks only a short time ahead?
8. Does period batch control produce items Just-In-Time?
9. In what ways is period batch control similar to and different from MRP?
10. What is the similarity between the approach suggested to achieve set-up time savings on heavily loaded machines in PBC and Goldratt's 'drum'?

REFERENCES

Burbidge, J.L. (1970) The case against stock control, *Proceedings of the Second Annual Conference of the British Production and Inventory Control Society.*
Burbidge, J.L. (1975) *The Introduction of Group Technology,* Heinemann, London.

Burbidge, J.L. (1988) Operation scheduling with GT and PBC, *International Journal of Production Research*, **26**(3).

Burbidge, J.L. (1989) *Production Flow Analysis*, Oxford University Press, Oxford.

de la Pascua, G.P. (1992) *The specification of a computer aided production management system in a capital goods manufacturing company*, M. Phil., University of Plymouth.

de la Pascua, G.P., Hallihan, A.J., Maull, R.S. and Childe, S.J. (1992) The design of a control system for group technology manufacturing in the high variety, low volume capital goods industry, *Proceedings of the Sunderland Advanced Manufacturing Technology Conference*, Sunderland Polytechnic 27–30 April, 1992.

Peters, T.J. and Waterman, R.H. Jr (1982) *In Search of Excellence: Lessons from America's Best-run Companies*, Harper & Row, New York.

Shingo, S. (1985) *A Revolution in Manufacturing – The SMED System*, Productivity Press, Cambridge MA, USA.

The development of a new CAPM system 10

Different forms of CAPM system have been described in earlier chapters. The difficult part is to make changes to improve real CAPM systems controlling real companies. This chapter looks at some of the theory about systems and describes some approaches to the design of a CAPM system to suit a company's requirements. The chapter is written from the point of view that successful implementation depends upon successful design.

10.1 SYSTEMS THEORY AND THE SYSTEM LIFE CYCLE

Systems theory provides a set of concepts which are of great value in the analysis of organizations. Systems theory has underpinned the explanations of the different types of CAPM system earlier in the book. The principles of systems thinking are useful in approaching the problem of designing a CAPM system. In particular, it is useful to understand systems concepts and the system life cycle.

10.1.1 SYSTEMS CONCEPTS

Systems concepts are the ideas which constitute systems theory. They provide a framework in which the world can be understood in terms of systems which are complete in themselves and which interact with each other. By identifying systems in the manufacturing organization, the interactions between the systems can be identified and the effects of one upon another can be understood, so that problems can be dealt with in an analytical manner.

A *system* can be described simply as 'a recognisable whole which consists of a set of inter-dependent parts' (Carter *et al.*, 1984). A change in a part of a system can be thought of as creating a new system.

The *boundary* of a system is a concept which allows the system or systems in question to be identified. It is common to identify boundaries around individual departments or divisions within the company which define what is in the department and what is outside. However, system boundaries do not necessarily conform to organizational boundaries. The 'order flow system' may identify a set of activities and people involved in processing orders, which may include people from many different departments. Identification of the boundary allows a certain set of people and/or things to be given a name and dealt with as a unit. It also provides a reference which can be used to identify materials or information flowing into the system as inputs and those flowing out as outputs.

The *environment* within which the system operates is defined by the boundary, since the environment is everything outside the boundary, but in practice it is most useful to consider the parts of the environment which influence the system or are influenced by it. Thus from a strategic point of view the environment within which a company operates consists of the stakeholders who have an interest in the company (see Chapter 1). A marketing perspective may see the company's environment in terms of its competitors, while a purchasing perspective might be to see the company in relation to its suppliers. From an ecological point of view the environment may be seen as the source of materials and the final destination of products and waste.

Every system within a business can be thought of as having *objectives*. A good understanding of the objectives of a system can help in improving and developing it. An example may be the quality department of a manufacturing company. If its objective is defined as 'to inspect goods', it may not concern itself with identifying the causes of quality problems in manufacturing. An objective related to improving the quality of the products as perceived by the customer may produce a more useful department.

Management appears in systems in the form of some kind of mechanism which regulates the activities of the system to help it to achieve its objective. Management may take the form of a set of activities which monitor the system's performance against a plan, such as the activities of ensuring that the correct orders are being worked upon at any time. Management appears in a *kanban* system when an operator stops work because no card is present, and when a card arrives and work starts again.

The sequence of activities performed by a system is often referred to as a *process*, especially where it represents a regular series of activities such as those concerned with accepting a customer order. (It should be noted that the use of the term 'process' here is distinctly different from the use in the group technology literature described in Chapter 7). Different processes can be used to achieve the same objectives. One company may use an MRP system to determine its purchase orders, while another may use a manual reorder point system. The design of new systems is often concerned with changing the process by which certain objectives are met.

The concept of *components* allows us to see systems as being made up of sets of things which may themselves be regarded as systems. This allows the same techniques to be used to identify the objectives, boundaries and interactions of the parts of a company as well as of whole companies. Thus work groups can be identified and dealt with instead of departments, and CAPM systems can be identified as part of the manufacturing infrastructure (Chapter 1). If all the production resources of a company are dealt with as a complete manufacturing system, then the infrastructure elements and the production activities identified in Chapter 1 constitute the components. The component which regulates the flow of orders and materials can be thought of as the management element, that is, the CAPM system.

The *viewpoint* of the observer studying a system is also important. In studying a CAPM system, it may be important to understand the activities of people who process orders, so that new ways of processing the orders can be found. In this case, flow charts may be used which describe the sequence of activities. Alternatively, it may be felt that the storage and flow of information relating to orders is more important to understand than the activities, in which case data flow diagrams and entity relationship diagrams used for designing databases may be thought more appropriate. Diagrams used to describe and represent systems are often known as *models*.

10.1.2   THE SYSTEM LIFE CYCLE

Systems change. Systems theory is a theory of change, which sees new ways of operating being created as old configurations need adjustment. As a business develops in a changing environment, its systems must continually be adapted. The system life cycle is used to identify the stages in the development of a system.

As was mentioned above, a change in a system may be regarded as the creation of a new system. However, since systems can be identified at any level of detail, from a general description of a company down to a very detailed account of a particular stage of a particular machining operation, the degree of change which can usefully be described as the creation of a new system depends upon the level of detail in the analysis of the system in question. If the production management system is being considered, a change in the information presented on a works order or *kanban* card may be considered a minor improvement. If the change is the replacement of reorder point logic by MRP then it may be considered the creation of a new system. The system life cycle therefore allows for both the creation of new systems and their minor modification.

It is usually the case that a new system replaces an existing one. The functions of production management may have been performed by entirely manual means in the past, and computerized assistance may be considered. Only when a new company is established, or when it must perform entirely new functions which it has never done before, is there no system in existence. In the design of new systems it is always useful to learn the reasons for the old system to work the way it does, since the experience the company has amassed in the past may well have led to ways of working for which the logic is not obvious. A new system should take into account the experience which is embedded in the old system, unless the company's experience is to be thrown away.

There is a counter argument that if too much is understood about the existing system it is too easy to lose sight of the system's objectives and to become convinced that the old way of working is the only way. In particular, where ways of working have grown and developed over a long period the objectives and boundary of the system may have been forgotten.

The system life cycle deals with these problems by establishing the conceptual basis for the new system and evaluating its feasibility in terms of its main objective. If it is clear that a new system can be beneficial, then further analysis will be used to develop a good understanding of the work situation based upon the experience embedded in the old system before designing the new system.

The rest of this chapter is structured around a description of the life cycle of a system in six stages: feasibility study, analysis, design, selection, implementation and maintenance. These six stages are meant to represent the changing activities in the life of a system, but should not be thought of as a rigid methodology for systems interventions. While the six stages do follow on in a reasonably logical manner, there is no intention to suggest rigid project planning which sees one stage completed and signed off before the next can start. Rather, a fluid, iterative process is to be encouraged which allows elements from the different stages to be used as required. Some design work may be considered in the feasibility study, and selection options may suggest further analysis etc. People are the most important parts of systems in business, and an iterative, flexible approach allows them to use their skills and to learn.

Feasibility study

The concept of the new system is evaluated. The idea of using a new type of CAPM may be examined on the basis of a clearly stated objective, such as an improvement in throughput or a reduction in inventory or cost. The way the improvement is to be achieved is considered from the points of view of the benefits expected, the cost and time taken to develop and implement the new system, and the skills required to make the changes. The benefits from the new system are evaluated as far as possible, so that the decision whether to proceed can be taken.

Analysis

Once a decision to proceed has been taken, a detailed examination of the performance expected of the new system is undertaken. Whereas the feasibility study will have considered the broad aims of the new system and a general approach to be taken, the analysis phase examines the proposal in detail with the aim of specifying exactly what performance is required of the new system in terms of its inputs and outputs and the processes which must be performed.

Design

The determination of how the new system will operate to achieve its specified objectives. Once an analysis phase has specified the role which must be fulfilled by the system, then design looks at how the role can be performed. Design involves the specification of the activities which are necessary for the system to operate.

Selection

The activities which have been designed must be performed by people and machines. The selection phase looks at the activities and determines which are the most suitable for execution by machine.

At the same time it is important to consider the computer hardware which is to be used, since the skills in installation and maintenance of a particular system will be required. It may also be important to use the same hardware type as is in use in other parts of the business. For example, it is often useful to integrate the CAPM system with the bill of materials file in the design system, if one exists.

Implementation

Implementation is the stage of the development of the system when the system is introduced into the company by installing the equipment, training the staff and beginning to use the new ways of working. The work of the implementation phase can be done alongside the other phases if a participative approach is taken – staff who have helped to design a system will already understand the new ways of working, although they may need to be instructed in the particular ways in which different activities are performed with the aid of a particular software package.

Maintenance and development

Over its operating life, the system is gradually developed by making small changes. As the people who use a system become more familiar with it, they may require minor changes to make it easier to use. Such minor changes might involve altering the layout of a printed report or the way information is presented or accessed on a screen. Maintenance changes may involve upgrading hardware to allow more storage space or faster operation, or to change to newer versions of the software to iron out problems.

More significant changes must not be carried out without taking steps to ensure that the systems objectives will still be met and that other related systems will not be affected. This means that a feasibility study of some sort should be conducted before making important changes. If the feasibility study is successful, the cycle turns again with a new analysis phase.

The phases of the life cycle will now be considered in more detail.

10.2   FEASIBILITY STUDY

Feasibility studies may be at any scale, from the discussion of alternatives between two engineers on a train journey to the commissioning of an investigation and report by a consultancy company. Between the extremes lie the various projects which companies may do for themselves and the involvement of outsiders, such as academics or government agencies which are established to advise businesses. Beyond these extremes, and beyond the world of manufacturing, Royal Commissions and Planning Enquiries may be thought of as types of feasibility study. The feasibility study determines whether or not to proceed with a particular course of action.

10.2.1   BOUNDARY SETTING

The feasibility study should set the boundary for the analysis, to determine which activities should be analysed and managed as a system. Current thinking advocates the study of systems defined around the flows of information or material through the business. This literature has described such systems as 'business processes' and has proposed their development through incremental improvement (Harrington, 1991) or radical redesign (Hammer, 1990; Hammer and Champy, 1993).

The process view sees the CAPM system as the system which controls the 'Order Flow Process', which converts customer requirements into products which are delivered and paid for (Weaver, 1995; Weaver *et al.*, 1995). In doing so the CAPM system brings together the management of activities connected with customer orders, such as master production scheduling, control of orders to the factory to produce products, including scheduling etc., and control of orders to suppliers for material. Other business processes involved with the operation of the manufacturing company include the development of designs from concepts, the activities needed to get orders, and the activities undertaken to support products after sales. These processes are dealt with by Maull *et al.* (1995).

The business process view encourages the company to see its processes through from start to finish, to ensure that the entire flow is dealt with. In CAPM, this means that attention should be given to customer orders from the moment they are taken until the customer is satisfied with the product and payment is received. Some analyses in the past have defined CAPM narrowly so that systems improvements have dealt with orders only once they pass from the contracts department to the factory, even though the customer may already have been waiting several weeks. Similarly, many systems pass responsibility to other departments when the product is despatched, and do not consider the activities required to ensure that the customer is satisfied and payment is collected. This results in companies failing to listen closely to customer feedback, and in companies continuing to deliver to customers who do not pay. The business processes approach helps to see the system as a whole from the viewpoint of the customer, even if the area which it is decided to improve is only a part of the whole process. It also encourages the removal of departmental boundaries within the company which delay the flow.

10.2.2 SYSTEM TYPE AND OBJECTIVES

The most obvious question to answer in the feasibility study for a new CAPM system is whether a system of a particular type, such as stock control, MRP, *kanban*, goal system or period batch control is most appropriate for the given manufacturing company.

For the large number of companies operating in batch manufacturing with a factory organized by type of machine and complex work flow, the most common answer is to use some form of MRP. This may be because most MRP-based systems try to take account of queue times and capacity problems while stock control systems do not, and The Goal System software and its predecessors are not well known despite the popularity of Goldratt's books. The other alternatives of *kanban* and PBC require the factory to be laid out in particular ways to support *kanban* lines or production cells, and often seem to be rejected for that reason.

If production managers could be encouraged to improve work flow in the factory before addressing the CAPM problem, it might be that great steps could be taken by simplification, making the factory much easier to control and removing the need for complicated computer systems. This is certainly the experience of companies who have successfully used *kanban* and group technology.

In establishing the feasibility study, it should be made clear that the option of simplifying the whole manufacturing system should be considered, and if work flow is a problem it should be dealt with before the control arrangements are considered. A simple way to show that work flow is a problem is to measure the proportion of the lead time of a typical component which is actually spent in useful operations. If the proportion of useful time spent is less than about 10%, then the simplification of the work flow should be considered a priority. At 10%, material spends 90% of its time waiting for something to happen, and inventory levels are about ten times higher than they need to be, but few batch production factories will beat 10% unless they are particularly well designed. A figure less than 2% would not be surprising in some traditional factories, where there is usually plenty of scope for improvement.

When the work flow arrangements have been considered, it may be necessary to investigate the improvements needed in the CAPM system. However, if the work flow has been arranged into flow lines or cells it may be very simple to move to a *kanban* or PBC system, and this may be designed in tandem with the work flow.

The objectives of the manufacturing system as a whole, the manufacturing strategy, must be considered. System types cannot be copied from one business to another – even in the same industry – if the companies have different competitive requirements. Different configurations of systems will be required to provide, for example, minimum lead time, minimum inventory, minimum product cost or instant delivery from stock.

10.2.3 CHOICE OF SYSTEM TYPE

De Toni *et al.* (1988) present a complicated discussion which aims to define fully the confusing terms 'push' and 'pull' in manufacturing.

Their paper can be used to help identify which CAPM variant is most appropriate for a given situation, although it is complicated by their interest in whether a particular control system should be classified as 'push', 'pull', or as a combination of both. The alternatives which they use as illustrations include MRP, reorder point stock control (ROP), *kanban* and synchro-MRP. Their analysis is based upon dividing the CAPM system into three areas:

- production planning (master production scheduling and assembly scheduling)
- inventory management (purchasing)
- priority assignment and picking and moving (shop floor control).

Master production scheduling and assembly scheduling

If manufacturing lead times are longer than the delivery times required by customers, master production scheduling must be to a forecast. If the assembly lead time is also longer than the customer can be expected to wait, assembly must also be to a forecast. These fall into the category of make to stock (MTS) of Wortmann (1989) (see Chapter 1).

If the assembly lead time is short enough, assembly can be to order while manufacturing of parts is to forecast (ATO).

If the total manufacturing and assembly time is short enough, or if lead time is not a major competitive dimension, production can be to order, as in both MTO and ETO companies.

Within the overall schedule, decisions must be taken to assign priority on resources on the shop floor.

Shop-floor control

De Toni *et al.* simply consider the degree to which manufacturing is repetitive in determining how to control the shop floor.

If manufacturing is not repetitive, a system which assigns priority on the basis of a simple rule, such as the critical ratio, is suggested. Critical ratios are priority rules based upon a ratio, such as number of operations remaining to be done divided by time available before due date. Such rules are little help in getting together a set of items which can be assembled together and assume that all items are equally important.

If manufacturing is repetitive, *kanban* should be used.

If manufacturing is semi-repetitive, synchro-MRP should be used.

Unfortunately, the degree of repetitiveness in manufacturing is impossible to measure usefully, since all industries see different levels of volume and variety as normal. It may be possible to re-state these rules thus. If *kanban* can be used, use it, or failing that, use synchro-MRP. If neither of these is useful, an alternative approach is required. Besides the critical ratio approach, alternatives are presented by Burbidge in period batch control and the rules for work sequencing within the period (Chapter 9), and by Goldratt in the subordination of all activities to the rate of production of the drum (Chapter 6).

Parnaby *et al.* (1987) suggest that one of the determining factors in the choice of system is the percentage of products of the three types:

Runners – Products which are regularly required and for which demand allows *kanban* style repetitive manufacture within a defined manufacturing planning period.

Repeaters – Products for which although there is a steady demand it is insufficient to economically allow repetitive manufacture within the defined manufacturing period ie products at the start or end phase of their life cycle.

Strangers – Products for which demand will occur but at low volumes and irregular frequencies eg prototypes, or aftermarket sales. Also products which do not have .... repeating similar features and are suitable for jobbing manufacture.

Parnaby *et al.* suggest that runners be controlled using a *kanban* control system with MRP for raw material provision. Strangers and repeaters can be produced as long as they are similar enough to use the same production line.

In contrast, Sandras (1989) believes that as long as parts which require the same operations are required, *kanban* systems can be used. He states:

What is important... is not that you repeat your product but that you repeat your processes

Purchasing

De Toni *et al.* point out that if the lead time required to purchase materials is longer than the time available, forecasting must be used to determine what to purchase. This means that ROP may be used, establishing the reorder point according to historical usage rates.

If the lead time to purchase materials is short, materials can be purchased for specific orders, such as by MRP, especially if the item is of a high value. If the item is of a low value, purchasing is simplified if reorder point logic is used, even though MRP is feasible.

This analysis is complicated by the fact that material items in the lower levels of a many-level bill of materials would be required much earlier than their own lead time, since before they are made into the final product they must be incorporated into higher level items. This means that the effective lead time is much longer, so that ROP is more attractive for these items.

De Toni *et al.* conclude that, from the JIT viewpoint, a marriage is desirable between MRP for purchasing, with safety stock and safety lead time removed, and *kanban* for shop-floor control.

All the discussions above are strongly influenced by the trend of production management literature of the late 1980s to deal with the relative merits of MRP and *kanban*, which was generally referred to as 'JIT'. The two writers which do not seem to have been swayed by this debate are Burbidge and Goldratt.

Burbidge's approach is one of simplification, to bring the complexity of the production system down to a level where it can be controlled by a group leader with some simple scheduling rules. Material purchasing in the PBC system can be by MRP or ROP, according to the level of variety. Burbidge places emphasis on the predictability of control and making the system resilient to disturbances.

Goldratt also takes a simplifying approach, although his logic is based around throughput as the key performance measure. Using his approach, all production activities are subordinated to the production rate of the constraint so that output is maximized while inventory and cost are minimized. Robustness of the schedule is also a key part of the system.

Both Goldratt and Burbidge present very convincing arguments which lead to robust CAPM systems that are simple in principle and which have built-in ways of dealing with variation. This makes them both superior to MRP for shop-floor control, which fails when lead times and production data are inaccurate (which they almost always are) and to the simpler *kanban* approaches which cannot tolerate fluctuating demand and a high variety of production.

Therefore, for the purposes of the feasibility study in most factories, it would seem useful to consider either theory of constraints or period batch control. It may even be possible to construct a cell-based TOC system which could be operated by a cell leader in the way that PBC can, but with the added advantages of identifying the constraints and the continual improvement which are part of Goldratt's approach.

If the factory under consideration has a narrow product range, the *kanban* control techniques should be considered, with either MRP purchasing or with *kanban* extended to suppliers according to the item. ROP may alternatively be used for the low-value items, but if it were to be used for high-value items it would remove the advantage of low inventory from the *kanban* system. If ordinary single-card *kanban* cannot cope with the required variety, the other forms of *kanban* could be considered as alternatives to TOC or PBC.

There seems to be no superior alternative to the use of MRP for the purchase of high-value items from suppliers, although safety stocks and safety lead time should be kept as low as possible. For cheaper items, and for items which have to be purchased to a forecast, there seems to be no superior alternative to some form of stock control system such as ROP or replenishment.

There appears to be no need for any factory to use MRP to control shop-floor production. This leads to all the well-known problems of MRP which stem from poor data, while superior alternatives exist.

10.3 ANALYSIS

The analysis phase may involve investigation of the existing system, not in order to copy it but in order to be able to understand what inputs and outputs are currently processed, so that they will not be overlooked, and to learn what pressures have resulted in the current system being as it is.

The focus of the analysis phase is more upon the environment into which the system must fit than the system itself. It is important that an improvement in one area should not cause problems in another area without good reason. Issues concerning the fit of the new system into its environment are often discovered at the design stage. It is useful, therefore, to manage the life cycle in a flexible way to allow iteration between the phases.

The identification of the system boundary and the use of modelling techniques to characterize activities and flows of material and information provide a structure for the analysis phase. The result of the analysis phase is a thorough understanding of the environment in which the new system must fit, and how it must interact with its environment. This may be documented in the form of models of the existing system and a specification of the requirements for the new system to fulfil. There will also be a considerable amount of knowledge in the heads of the people who have done the analysis. Not all of this information is easy to document, so it is a good idea to allow the same people to have some involvement in the design phase. For the same reason, it is useful to use staff who are likely to remain with the company, even if the expertise of outsiders is required to support available staff in techniques such as modelling.

At its simplest, analysis involves identifying the inputs and outputs which the system must process, in order to be able to determine exactly how the system must be designed to fit into the environment and deliver its objectives. The inputs and outputs of the current system need not be copied exactly, but it is important to identify them since it is by the inputs and outputs that the system fits in with the rest of the company. If they do not appear to be needed, or if other inputs and outputs appear more relevant, this must be studied in connection with the other systems in the company.

An example of a system which is commonly computerized is the part of CAPM which issues orders to suppliers. If such a system were to be computerized, analysis might determine:

the outputs:
- the information to be contained in each order
- the information which will be sent to the receiving dock to check the items on receipt
- the information to be stored about orders which have been raised
- the information to be stored about deliveries which have been made
  the inputs or the information which needs to be supplied to the system from other systems:
- the number of items
- the item number
- the requirement date.

Information will also be collected in the analysis phase concerning the volume of work involved, such as:

- the number of orders to be raised in a certain time period
- the number of items for which supplier details will be recorded
- the number of suppliers whose details must be recorded.

Although the process by which the inputs will be turned into outputs may be thought of as part of the design phase, the analysis phase may suggest some activities which the system might be expected to perform, since they currently appear to be necessary. These may include:

- authorize purchase of the item
- check or adjust the requirement date
- read from files details of the supplier for the item
- check there is an agreement in force with the supplier
- transmit the order to the supplier.

At the analysis phase it may be identified that some of the activities could be removed by designing the system in a certain way, such as making sure that the requirement date has been checked and the order has been authorized before it enters the system. It may also be suggested that there is a need for a certain activity to be performed in certain situations, and for certain activities to be performed by humans rather than computers.

The analysis information provides a specification which can be used in designing the system. If the system is a simple one, it may be sufficient to provide information such as that shown above in a report with suitable explanations.

In more complicated situations, it is usual to use graphical representations of the system. These are usually much easier to understand than verbal explanations and provide a way to show a large amount of information in a few words.

10.3.1 SYSTEM MODELS

System models are usually graphical representations of systems. Each model may be made up of diagrams or charts and explanatory descriptions. A range of standard modelling techniques has become well known across many industries, because they have been designed to allow staff from different backgrounds to communicate their understanding of a system which may concern them all. For example, in the design of a system it may be necessary to conduct an analysis phase involving design engineers, buyers, accountants, receiving dock staff, computer system designers etc. While the detailed design of elements of the system may require specialized modelling techniques, such as for the design of databases or computer networks, general techniques are very valuable in establishing the requirements of the system in the analysis phase.

What makes a good modelling technique?

A system model must show the characteristics of the system which are of interest in the analysis. If the activities performed in the system are to be analysed, they must form part of the model. If the information flowing in the system is to be examined, it must be shown. Similarly, characteristics such as authority, decision, material flow and information storage may all be of interest.

Often it is a main aim to reduce the cost of operating a system and the time taken for it to produce its output. Unfortunately, money and time turn out to be particularly difficult to model. These can be dealt with to some extent by addressing activities which take time and must be paid for, so cost and time information can be determined with the help of a model showing activities.

Models are used to help people to understand the system and to come to agreement about the current situation and the ways it could be improved. Models must therefore be easy to understand, although producing them is always more difficult than understanding them. The model should also be drawn at the correct level of detail for the analysis required. A model which is too abstract may be difficult to criticize yet of little use. A model which is too detailed may draw attention to trivial points while more important ones are hidden.

The support for a particular modelling technique must also be considered. Not all techniques are widely known, and training can be both time-consuming and expensive, although it must be considered a long-term investment if a particularly valuable approach is made available for future projects. Computerized tools are available to help in the drawing and editing of some types of model, for which the availability, cost and training must be considered. The question of producing multi-purpose models which show several aspects of a system, such as both activities and data requirements, is a topic which is currently being addressed by researchers and software companies.

Two common modelling techniques are:

- flow charts, which show the sequence of activities in a simple process
- $IDEF_0$ diagrams, which show activities together with the conditions controlling each activity, and which can be used to handle more complex situations at various levels of detail.

These are described in Appendices A and B.

Other modelling techniques

The techniques above both focus on the fundamental question of 'what happens?' although $IDEF_0$ also shows the circumstances surrounding each activity and the means of performing each activity, together with the essential information flows. Other techniques may be required to focus on other aspects of the system, such as the data concerned.

Data flow diagrams show the way data is stored, processed and transmitted to entities outside the system. Data flow diagrams, often known as DFDs, are a commonly-used technique of computer systems analysts. The DFD is a part of the Structured Systems Analysis and Design Method (SSADM) (Downs et al., 1992; Skidmore et al., 1993), although similar techniques are found in other standards.

Another data model is the entity relationship diagram. This shows the basic structure of the data which is used by a system, and is particularly suited to the design of databases. One standard entity relationship diagram is part of SSADM (Downs *et al.*, 1992; Skidmore *et al.*, 1993), while another is part of the IDEF suite of techniques, $IDEF_{1X}$ (LeClair, 1982).

10.3.2 GENERIC MODELS

The modelling of CAPM systems can be extremely time-consuming, particularly when the company is unfamiliar with modelling techniques. In order to reduce this problem, generic models have been proposed by the author (Childe, 1992), and developed and tested by Weaver (Maull *et al.*, 1995). The generic model approach has three aspects:

- It provides a structure which allows a company to analyse its own systems without the effort required to develop its own models from a blank sheet of paper.
- It allows deviations from the generic model to be explored, so that the company identifies which parts of its operations are unusual. These areas can then be concentrated on either removing unnecessary activities or understanding the company's particular strengths.
- It allows the company to form a consensus view in the light of a model which has no internal political significance.

The generic model developed by Weaver uses the $IDEF_0$ modelling technique and has been widely validated.

10.3.3 STATEMENT OF REQUIREMENTS

The outcome of the analysis stage should be a clear understanding of the requirements that must be fulfilled by the new CAPM system.

This should cover the functions of the new system and is often expressed in the form of an outline model or a document describing the required functions in the areas of order processing, planning and scheduling, purchasing, stock control, shop floor control, delivery arrangements, payment collection arrangements etc. It may cover any or all of these areas or may take a different structure according to what seems appropriate for the particular company.

Even with the use of generic models, each company must determine its own solutions, although there will always be similarities at the more general level. Where models such as $IDEF_0$ are used, the requirements statement may consist of a model with a structure but without detail in certain areas, the detail remaining to be determined by the design phase.

The structure of $IDEF_0$ lends itself to this way of specifying requirements, since it allows an activity to be specified in terms of its relationship with other activities but without necessarily identifying what will go in the box at a more detailed level.

The statement of requirements is a useful marker to show the transition from the analysis to the design phase. In many traditional approaches it provides the opportunity to sign off a report and hand it over as a set of firm requirements to a design team who will determine how to fulfil the requirements. A less rigid approach may see the system life cycle as a set of phases which describe, but do not constrain, activities, when it may be permissible or desirable to reiterate between the phases, particularly between analysis and design. In this case the specification of requirements may be expressed as the state of a set of models and descriptions at the point when attention shifts from examining the existing arrangements to defining the new arrangements. This point may not be clear at the time, especially when the same team follows the project through both the analysis and design stages.

The transition between analysis and design may be much less distinct than the following white space suggests.

10.4 DESIGN

The design phase benefits from a good level of creativity and a readiness to consider doing things in new ways. Together with staff who understand the problem, it is often useful to have the involvement of people who do not share taken-for-granted assumptions and who can ask questions and help create solutions which have not been used before. By the use of creative design teams the company can reconfigure itself to operate in a way which is different from its competitors.

It is useful if design can involve the staff who will be expected to work in the new system, since they will then understand why the new system is better than the old one, and they can take part in decisions which will affect the way they spend their working time. Decisions taken at too high a level by people who do not have to operate the system may cause unnecessary inconvenience and resentment.

The design stage should also determine what information will be required to perform each activity. By considering the flow of information at the same time as the sequence of activities it is possible to make improvements to the flow. This is broadly similar to the way in which group technology designs production cells in order to allow whole components or products to be produced at once. It may be possible to design the information system and activity system in such a way as to allow a data item such as an order to be processed fully in one activity rather than in stages.

The result of the design phase may be a set of models which describes the functions that will be performed by the system to a sufficient level of detail to show the sequence of activities involved.

10.4.1 THE DESIGN PROCESS

A simple way to achieve new designs is to produce a system diagram showing the inputs, controls and outputs of the required system (the statement of requirements) and using a group workshop to generate ideas on how the work could be done. A simple comparison technique could be used to evaluate the best option, and further design work could be used with the group of people concerned with the system to produce a workable solution. This approach may be seen in three phases as:

- collect alternative ideas
- filter the ideas
- refine the design.

Alternatively, a team may seize upon a preferred solution and defend and advance it without seriously considering a range of alternatives, because they all share a common consensus on the way the thing should be done. If that is the case, there is little point trying to force the design process or any part of the system life cycle into a rigid procedure. The aim is to find the consensus solution, irrespective of how it is produced.

Design is an intuitive, creative process, and the only way to create the new system design is to use whatever skills are available in the way which seems to be the best. There is no best way to produce the new design, because the best new way might just be a way that no-one has thought of before. There is no way of knowing that a particular idea is the best, except that everyone concerned agrees it is the best. If that is the case, then it is the best you can do.

The design process relies upon inventive thinking, which in turn means that the people involved must be in a positive frame of mind, and willing to contribute new ideas. In order to facilitate new designs, it is almost always taken for granted that a team approach will provide the arena in which new ideas can be generated. Of course, new ideas can come from inspired individuals at odd moments, but these are hard to arrange. At least if several people reflect on a particular problem, the likelihood of one of them producing a good idea is increased.

Various ways of stimulating the creative process have been published, which mostly address the generation of alternative ideas.

However, it is equally important that the evaluation stage should not throw away unusual ideas which may prove to contain the best solutions. Design refinement looks at the polishing of the idea into a useful form and may involve taking a strange idea and proving its worth in a pilot scheme, model or prototype.

10.4.2 IDEA GENERATION

Ideas may be generated at any time, some of which may actually start a new turn of the cycle or at least an improvement of the existing system. Various techniques exist to encourage creative thinking in design teams. The most well-known is 'brainstorming', in which a group attempts to create as many ideas as possible of ways to solve the particular problem. Brainstorming relies on the rule that no idea must be questioned, but simply recorded. The most far-fetched ideas may trigger other thoughts, so that after a few minutes a board can be covered with ideas. Brainstorming is described by Osborne (1963).

Without a detailed understanding of how people actually do think, it is difficult to see how people can be made to think in new ways, but at least it is possible to show people that their existing way of thinking has become habitual and regularized, and help them to question existing ways of working and of solving problems.

A colleague uses a simple demonstration to show people how their thinking is regularized. At the start of the session, everyone in the room is provided with a banana and told it is to be used in a creativity exercise later. At the appropriate time, the speaker asks everyone to unpeel their banana. It is then pointed out that everyone uses the same method, that is, to bend the stalk to break open the peel. No-one ever uses a knife to cut the peel, and no-one ever opens the fruit from the end where the flower grew. The point is made that people just go on doing things in the ways that they have always done them, without questioning whether those ways are the best.

Unfortunately, even once the point has been made, it is not a reliable assumption that this example will help people become more creative. They may start to question the way they peel fruit, but they may not be able to produce any better way of determining when to release orders to suppliers. They may allow themselves to question things from time to time, or they may just switch to always peeling a banana from the blunt end.

Another example uses a puzzle with matches. The willing volunteers are each provided with nine matches and asked to make three equilateral triangles. The task is easy. Three matches are then taken away, and they are now asked to make four equilateral triangles with six matches. This is a good challenge for business people because they appreciate being asked to make more with less. Some people will immediately get they idea of using a third dimension, to produce a three-dimensional tetrahedron. Others may start breaking the matches into pieces. Either way shows a spark of creativity, since many people will make the first triangles with whole matches lying flat on the table and will assume that this is the way the second challenge is to be done. Creative thinking is praised, but whether everyone is made more likely to think creatively in future is still in question.

Some writers suggest that individuals can be taught to think more creatively. Good examples are to be found in von Oech's 'A Whack on the Side of the Head' (1983), the work of Edward de Bono on lateral thinking (for example de Bono (1995)), or Lumsdaine and Lumsdaine's collection of techniques (1995).

Training is vital to the success of any company which employs people, and training in creative thinking is an important example. However, when engaging on any particular project such as designing a new CAPM system, there is unlikely to be much time for people to train, and ways have to be found to gain the best from the individuals available at the time.

If individuals have only a limited amount of creativity, it makes sense to use teams for creative design, so that team members with different points of view can stimulate new ideas in others and question their assumptions. It is particularly important not to allow groups of people who are used to working together and who share the same opinions and assumptions to work together in a creative team without at least having an outspoken 'Devil's advocate' to question their ideas. A better approach may be to use teams composed of people from various different areas concerned with the problem but who do not work closely together, with other people who are inexperienced and are willing to question. Such candidates may include trainees and apprentices, customers, suppliers, friends and neighbours, and possibly even academics and consultants.

Some new design ideas may be drawn from the literature. Harrington (1991) proposes an approach aimed at the removal of bureaucratic procedures and non-value-adding activities. Another view of the problem comes from Jain *et al.* (1994), who propose a series of questions based on the activities identified in the analysis phase. These include:

- *Can an initiating task be moved to the customer?* Some companies have made savings from allowing customers to design products from modules in a catalogue, thus removing some design work and simplifying the entry of orders. It is common nowadays to dial your own telephone number, but in the past this job was performed by telephone operators. A technological development, the automatic exchange, made this change possible. There may be little gain if a task is moved to a customer department within the company, as the company must still do the work, unless it requires less effort to perform the task in a different area.

- *Can any task be moved to the customer?* Some customers may be willing to arrange to collect their own products from the factory rather than having them delivered. Others may even be happy to finish the manufacture of the product – consider the advantages to the furniture company who first thought of packing furniture in flat-pack kit form.
- *Can activities be combined?* For example, it is common for someone to check stock levels and write a note to the purchasing department to ask for purchase of the required items. These activities could be combined if the same person was able to do both parts of the job.
- *Can activities be performed at the same time?* It is common to think of the flow of information as being linear, one activity at a time, but if opportunities can be found to perform activities in parallel it can make a great saving in time. An example of this is in close scheduling, where a batch of items are transferred from machine to machine as individual parts, so that a series of machines are all working on the same batch at once.

10.4.3 EVALUATION OF ALTERNATIVES

Having created a list of possible solutions to the problem, which may include detailed proposals and vague ideas, the next stage is to evaluate which alternatives should be pursued further. Like the whole of the system life cycle, this can be taken step by step or in an entirely unstructured way, as decided by the people concerned.

A structured way of evaluating alternatives is to determine (by whatever means) the absolute constraints which rule out some options and the criteria which can be used to compare the remaining alternatives. The alternatives can be filtered against the constraints and the remaining ones can then be compared. Constraints may include technical requirements such as 'must not require a new computer to be purchased' or 'must provide the following information...'.

The 'paired comparisons' technique provides an extremely rational way of sorting out alternatives and criteria, which despite its complexity may help if there are a lot of alternatives and criteria. However, it is more likely to help by providing a basis for discussion than by the solution it generates. This technique is described in Appendix C.

10.4.4 DESIGN REFINEMENT

Further investigation of solution alternatives can proceed in various different ways. It is important that all parties concerned with the new system gain the best possible understanding of how it might operate, so that they will recognize the need for changes and the way the changes affect their own work. When the decision is taken to proceed in a particular direction they should recognize that the best decision has been taken.

Models of the proposed new ways of working can be drawn up using a suitable modelling technique. This may simply mean extending or modifying the models drawn up during the analysis phase. The models should show the new activities which people will be called upon to perform, and should show what information flows will be necessary between individuals and groups to make the system work. They should show what people must do, how it must be done and who it must be done with.

Simulation may be required to show that a particular system will be able to operate under changing conditions. Computerized simulation systems are able to show how queues of work may build up if various fluctuations in workload occur.

Prototype versions of software and documents may be created so that different aspects of the new system can be examined. Presentation can be tested to ensure that all the required information is shown in the right place in the right way. Computer packages can be tried out to ensure that their operation is understood.

Pilot schemes may be used to introduce the new system in a restricted area, so that it can be tested for real before full implementation.

A successful design phase should produce a design of the new system which is well understood by the staff concerned and which has been thought through and tested as far as possible. However, the boundary between the design phase and the following two phases is unclear in many instances.

10.5 SELECTION OF OPTIONS TO FULFIL THE DESIGN

In the CAPM area, this means using a computer to do repetitive tasks such as the gross-to-net calculation which determines the net manufacturing and purchasing requirements to fulfil a set of orders. The activities which must be performed by humans include those activities which involve important decisions, such as setting the master production schedule. These activities require information which it may be most appropriate to use the computer to store and manipulate. There are also some tasks which fall to humans as the penalty for using computers. These are the tasks which are needed to allow the computer to operate, such as preparing and keying in data, and performing maintenance tasks such as the backing-up of data files for security.

For every task which is to be computerized, an appropriate piece of computer software must be available. The two alternatives are to produce software using a programming language or to purchase a software package which provides the required functions.

A wide range of tried and tested software packages is available, many of which perform activities related to MRP and reorder point stock control. It is not usually worthwhile for a manufacturing company to try to become a software house to develop a software solution which could be purchased. It is not uncommon, however, to develop extra applications to suit special requirements of the particular company.

In selecting a package, it is important to determine which extra tasks will be required in order to run the software, in addition to those which the software is meant to perform. These extra tasks must be performed by the people concerned. At the same time, it is likely that no software package will perform all the tasks which are identified as suitable for computerization. This means that the best design is a compromise which can be found by considering several alternatives and the effect they have in generating manual tasks. It may be necessary to perform several iterations of the selection process until a set of human tasks is arrived at which is felt to be acceptable together with a set of computerized tasks which together perform the functions required of the computer and which can be obtained as a package.

10.5.1   THE USE OF PEOPLE AND COMPUTERS

The main decision in determining how any task in the CAPM system should be performed is whether or not to use a computer.

March and Simon (1958) look at the nature of the work to be performed in order to deal with the question of whether it is appropriate to allocate tasks of various types either to people or to machines. Their findings can be applied in the question of whether to computerize a particular task, according to the extent to which the decisions involved in the execution of the task are 'programmed'.

Decisions are programmed to the extent that they are repetitive and routine, to the extent that a definite procedure has been worked out for handling them... if a particular problem recurs often enough, a routine procedure will usually be worked out for solving it....

Decisions are nonprogrammed to the extent that they are novel, unstructured, and consequential. There is no cut-and-dried method for handling the problem because it hasn't arisen before, or because its precise nature and structure are elusive or complex, or because it is so important that it deserves a custom-tailored treatment. (Simon, 1960)

At one extreme the individual has to make a 'search' for alternatives, or create new solutions to the problem, whereas at the other extreme the decision has been well rehearsed and the appropriate response is clear. When the problem is new to the individual, no routine exists to deal with it, so the individual must search for and evaluate alternative courses of action. When the individual applies creativity to a problem, the search for alternatives is also the development of a routine to deal with the problem when next it arises.

For information processing activities (which includes those of production management), the computer may be able to perform the tasks which are 'programmed' to the extent that the constituent elements of the program can be made available, that is, expressible in a useful form, and used to make a suitable program for the computer. One difficulty when using computers is that the expertise which has built up in the heads of individuals is very difficult to formalize and capture in a computer system. Experienced people very often have ways of dealing with unusual problems which cannot be made accessible without a thorough understanding of the existing system. This is a strong reason why teams of people concerned should be used actively to design the new system.

Programmed decisions include the many calculations which are made in CAPM, such as the gross-to-net calculation used to determine purchasing and manufacturing requirements, and the decision whether to make an order based upon a reorder point. Such a program is described by March and Simon (1958):

1. When material is drawn from stock, note whether the quantity that remains equals or exceeds the buffer stock. If not:
2. Write a purchase order for the specified order quantity.

In the execution of such a program, no creative thought or discretion is required.

Non-programmed decisions require a search for alternatives, which may require the use of creativity and discretion. Non-programmed tasks include those which deal with customers and suppliers, such as when to schedule a large order or how to get hold of urgent materials, and more strategic decisions such as whether to make or buy a certain item. The decisions taken in the design of the CAPM itself are also non-programmed, since they are not done routinely and they are very consequential for the company.

Jordan (1963) pointed out that the human characteristics of flexibility and adaptability make humans fundamentally different from machines. Humans must not be thought of as being like machines with different specifications. Systems should be designed with the abilities of the human operator in mind, producing a design in which human functions are made use of, not subjugated to acting as part of the machine.

The use of human skills depends upon the nature of the skills available, and the skill of the system design team to create jobs which contribute to the overall system. A set of graded tasks may be provided to allow different people to use their different skills; to provide job rotation or autonomous working groups; to provide a route for personal development, promotion or training. Most importantly, the design of the system must result in meaningful, whole jobs to make the best use of people and to provide an acceptable working pattern. This aspect is discussed by (for example) Miller and Rice (1967) and Leavitt (1962).

Unfortunately, there is no simple rule for the choice of human or computer means for a production management task. A simple guideline would be to consider allocating programmable tasks to computers, while leaving the creative and 'search' tasks to humans. Ultimately, however, it seems that day-to-day pressures on managers are likely to result in computers being given the tasks for which computer programs are most readily available. It is to be expected that most CAPM systems will include computers to perform a selection of the programmable tasks, and that the rest of the programmable tasks and the task of supporting the computers, will remain with humans.

Broedner (1985) observes that:

> Most managers and production planners follow a strategy to replace human work... by enforced use of computers.... Since this strategy is in danger of creating new problems, the growing minority seeks to avoid them by reorganizing production and rearranging the division of functions between man and machine in a way that makes use of workers' skills instead of reducing them to operating servants.

Humans are more adaptable to change, and are especially good at making improvements to working practices, exerting judgement, changing from job to job, coordinating operations between individuals, dealing with variable input (information or material), correcting errors, and to some extent to changing output requirements (doing new things and accepting impossible demands).

The fantastic abilities of people to plan, remember, and use judgement, wisdom and intelligence extend far beyond the capabilities of computers and mechanization (Skinner 1985)

Humans are less good at coping with extreme repetitiveness (such as high volumes of data, long production runs), maintaining consistent output over long periods, and coping with large numbers of input sources.

Computerized solutions are better at dealing with vastly repetitive tasks and coping with large data volumes and numbers of inputs, as long as the circumstances do not exceed the limits which were taken into account by the programmer.

Computerized solutions are less good at responding to changing output requirements, detecting irrationality and 'thinking'.

In situations of repetitiveness, such as where high volumes of data are involved, computerized solutions should be considered. In situations where frequent changes and new situations must be dealt with, manual approaches may be more successful.

10.6   IMPLEMENTATION

It is important not to underestimate the difficulties of implementation. While installation is well understood and mostly predictable, implementation involves managing the expectations and activities of people, which cannot be so simply understood and managed. Implementation is not complete until the system is brought into full use to the point when the company begins to benefit from it, although the success of the implementation may depend upon all the stages from the initial concept of the system onwards.

Much work has been devoted to the successful implementation of computer systems and other systems using new technologies. A key theme in this work is the need to simplify the whole system before automating a part of it. This means that the key to successful implementation lies in the feasibility, analysis and design stages. If the system in question is seen as part of a bigger system involving other technologies and people, and steps are taken to integrate the system with its environment, the implementation difficulties of securing commitment and acceptance are much reduced.

The success of the implementation of a new system depends enormously upon the design of the system. If a team of people work on improvements to their own area of the business, it is hard to imagine that they would not take the step of implementing the changes that they had decided were necessary. Unfortunately, however, the question of implementation continues to be a problem, and many systems which are put in place are classed as failures of implementation because the business was not able to derive the expected benefit from the change. Successful implementation (as opposed to installation) means that the whole system, both computerized and manual elements, has been put in place and made to work so that benefits can be measured. Implementation involves fitting the new system into its context so that its outputs are useful and its inputs are available.

Successful implementation depends on the success of all the previous stages of the system life cycle, since a wrong decision at any stage would lead to a system which was not entirely suitable. Implementation should not be seen as a stage which comes at the end of the process, but rather it should be thought of from the start.

10.6.1 IMPLEMENTATION FAILURE TYPES

In a review of the literature on CAPM successes and failures, Meredith (1981) presented a vast number of references to surveys reporting rates of failure in computer system implementation of up to 99.7%. The proportion of implementations seen as failures by the various authors is so large that the implementation problem has to be taken seriously.

There is still a steady stream of literature on the failure of information systems implementations.

At the time of Meredith's review, personal computers had not yet become widespread. Most of the systems concerned involved mainframe computers, which tended to be specified and installed by computer specialists rather than by people with a good understanding of the business concerned. This must have led to the systems being installed with little involvement from the users, so there is likely to have been very little mutual understanding of the details of individuals' jobs and their combined experience. However, the classes of failure which Meredith draws out are still relevant, since they turn out not to be connected with the particular type of computer system in question. Meredith concluded that after twenty years of computer systems, the reasons for failure were not well understood. He put forward three distinct groups of factors which affect implementation failure and success. The three groups are technical factors, process factors and inner-environmental factors.

Technical factors

These include the basic facts about the system and its context which can cause failure irrespective of how carefully managed the project may be in other respects. They may be regarded as simple prerequisites to implementation success.

Data must be available to the system, and it must be correct. It is commonly stated that a high level of data accuracy is required for MRP implementation, and if a system fails because the data is too inaccurate it would be classed a technical failure. If data is not available in the correct form, the system must be designed to include the function of data preparation. Here it can be seen that one of the simplest causes of implementation failure depends upon correct design.

The next factor is simplicity, which is similarly a result of good analysis and design. Simplicity is cited by many authors as a requirement for success. Meredith points out that this is a problem for system vendors, since computer packages have to be made more comprehensive than any user requires, so that they can be sold to a wide range of customers. Flexibility to adapt and develop the system may be more easily available in a more sophisticated package, but this is likely to be more expensive.

Training is cited as a vital technical requirement, and it should be continuous and ongoing, with provision for new staff, rather than a one-off experience.

The implementation team should include representatives from all areas affected by the system, and should include a project team manger reporting direct to the head of the organization.

Implementation must be recognized as an important aspect of the overall project, not a subsidiary activity after design and selection. Implementation should also be regarded as an ongoing task with continued attention to allow the system to evolve.

Process factors

These factors are concerned with the way the project is managed, and may be thought of as questions of style. These are less easy to define.

A key factor is the involvement of top management. Support for the project is essential, but full attention is better. Best of all is the case where the project is initiated by top managers.

User involvement in the design of the new system is the only factor consistently correlated with implementation success. Meredith uses the term 'user' – as many writers do – in a way which implies that the (computer) system serves those individuals who work with it. This view encourages people to become involved with the design of the computer tools they use, but differs from the author's view that the system should be seen to include both the people concerned and the computers they use (if any). User involvement does not go far enough, since it presupposes that it is the computers which change while the activities of people remain basically the same. If the people are part of the system, they must become responsible for achieving the system's objectives and for the design of their own jobs as part of the system.

Inner-environmental factors

Most abstract of the three groups is the group of inner-environmental factors. These are attributes of the situation within the company into which the system must fit. (Meredith uses the term 'inner' to make it clear that he is referring to the environment within the company – the environment of the system not of the company.)

It is simple to assume that the system to be implemented must be important to the organization, but it is an important factor that the need must be real and must be felt by the important characters in the organization whose real support may be needed to allow investment of time and effort in the new system. Ideas which are experimental are seen as much more likely to fail, because if there are any problems the implementation could be dropped. Similarly, the enforcement of a 'not-invented-here' solution is much more difficult. This factor could be described as 'the need for the new system to be needed' or 'the importance of the new system being important'.

Managers and other opinion formers must be willing to change from the old system to the new, and must ensure that the existing ways of managing do not encourage people to hang on to the old way of working. This is dealt with in various contexts in a paper by Kerr (1975). This problem can arise when performance measures are applied to a characteristic of the system rather than to its objectives. An example is where purchasing staff are measured by the number of orders they raise, not by the availability of material or the (low) stock level – it may not be very easy to get them to establish call-off procedures which rely on one order per supplier per year. Similarly, if a new system is to become part of the way of life in the company, it is important that opinion-formers continue to use it. The manager who reverts to the old way in a crisis signals to others that the new way is not to be taken too seriously.

Willingness and commitment to change may also be called for in the way the new system changes working relationships between different departments or groups in the organization. Existing management structures may inhibit the performance of the system by splitting a single business process so that parts of the activity happen in different areas and information may be lost or delayed, or priorities may change at the interface between the departments. Concomitant to this is the tendency to redesign only the parts of a system which fall within a particular organizational area. The extent to which the organizational structure spoils the way the business operates can often be understood at the design stage by identifying responsibilities when models of the new system are produced.

10.6.2 METHODOLOGY

Before implementing any solution, Meredith suggests a checklist approach which will ensure that the causes of failure are not present. First, he suggests checking the environment by asking a series of questions of top managers along the following lines: Is this a crucial opportunity or problem? Is this clearly the only or best alternative? Must it be done now? Are significant resources committed? Is the gain expected to be substantial and measurable? Are we willing to change the organizational structure? Will we actively use the new system? Are we willing to accept changes to the basic way we operate? Does the new system fit the organization, or are other changes necessary?

Next, the process factors should be addressed. If top management is written into the project plan, participation of all the people concerned must be ensured.

Finally, the solution should be examined in the light of the technical factors to ensure that none of the easy issues has been neglected.

This approach to implementation operates mainly on the negative side, by attempting to remove the likely causes of failure. They should be addressed at the start of the project, rather than just before installation, since problems may be caused at the early stages in the analysis, design or selection phases.

On the more positive side, steps can be taken to ensure that the CAPM system fits the needs of the business. If the need for a new system is clear, it is much easier to facilitate the development of the new system. This can be addressed by re-evaluating the company's manufacturing strategy. If the aims of the business are clear, the performance of the CAPM system can be evaluated in the light of those aims, and any shortcomings can be addressed in the development of the new system.

10.7 MAINTENANCE AND FURTHER DEVELOPMENT

Once implemented, the new system must be allowed to develop and evolve. Manufacturing is a fast-changing business, and all systems need to be continually developed so that they offer new capabilities to maintain the company's competitive advantage.

In the early days of the life of a new system, it may be judged appropriate to prohibit any changes at all until everyone concerned is happy that the system is running as expected. Only changes which are needed to correct important problems should be allowed, so that attention is focused on achieving correct operation.

Once the system is fully operational, minor changes can be considered to simplify and improve the operation of the system. It is important not to neglect the system after initial implementation, when there may be great pressure to move on to the next project. Minor changes to the information held on forms, classification systems, computer instructions and menus can make a big difference to the life of the people in the system. After securing their cooperation to help design, implement and operate the new system it is important to continue to listen to the improvements they are able to suggest.

People who feel they are part of the system should still be listened to long after the system is no longer new, and they should be encouraged to participate in ongoing discussions about the way the system operates. Models of the system should be available so that minor changes can be worked out by the people concerned as and when necessary to deal with minor changes in the business.

Careful attention to the views of the people in the system can result in a human or human–computer system which is capable of something computer systems cannot yet manage – it can redesign itself as needed. This can be facilitated by encouraging people to review their methods through team approaches and providing ongoing training and contact with other companies, especially customers and suppliers, so that they are often exposed to new ideas to challenge the existing way of working. Ideas generated in the company must always be listened to, as they may provide the company with its next step forward in competitiveness. When an idea is found which calls for a radical change to the system, the life cycle begins to turn once again. If improvement becomes continuous, the cycle is always turning and all the phases start to merge into a continuous system development activity.

SUMMARY

This chapter has introduced some of the principles of system design as applied to CAPM systems. A CAPM system can be thought of as a set of people, routines and activities which manage the flow of material and information to satisfy customer orders. The concepts of systems theory provide a basis for the analysis of such systems, and the system life cycle provides an outline structure for the stages of the development of a system.

The phases of the life cycle, feasibility, analysis, design, selection, implementation and maintenance may be seen as a logical progression but in most cases the distinction between the phases is blurred by re-iteration and the combination of phases, such as elements of design which are done at the analysis stage and elements of selection which become part of the design phase. The implementation phase may be seen as encompassing the whole cycle, since its success depends upon all the previous phases.

Whether in distinct phases or continuously, the manufacturing manager must always consider whether existing systems are able to provide the best performance to satisfy the company's strategic requirements, whether the type of CAPM system in use is most appropriate, whether the system has drawbacks which need attention, how improvements should be made, how to make the best use of humans and computers, how to have the best chance of success in implementing changes and how to ensure that systems continually update and renew themselves. This is the challenge of production management.

QUESTIONS FOR DISCUSSION

1. What is a system?
2. Explain the terms boundary, environment, objective and viewpoint.
3. Why is consensus important?
4. What are the six phases of the system life cycle? What is the purpose of each phase?
5. Why are models used in the analysis and design of systems?
6. How can people be encouraged to be creative?
7. Who should be involved in system design?
8. What is the difference between 'implementation' and 'installation'?
9. Why is the life of a system described as a cycle?

REFERENCES

Broedner, P. (1985) Skill-based production: the superior concept to the unmanned factory, in *Towards the Factory of the Future* (eds H.-J. Bullinger and H.J. Warnecke), Springer, London.

Childe, S.J. (1992) The use of a generic task model for the design of computer aided production management systems, *Proceedings of the Sunderland Advanced Manufacturing Technology Conference*, Sunderland Polytechnic, April.

Carter, R., Martin, J., Mayblin, B. and Munday, M. (1984) *Systems, Management and Change*, Open University/Harper & Row, London (reprinted 1988 by Paul Chapman Publishing Ltd, London).

de Bono, E. (1995) *Serious Creativity*, HarperCollins, New York.

de Toni, A., Caputo, M. and Vinelli, A. (1988) Production management techniques: push–pull classification and application conditions, *International Journal of Operations and Production Management*, 8(2), 35–51.

Downs, E., Clare, P. and Coe, I. (1992) *Structured Systems Analysis and Design Method*, Prentice-Hall, Englewood Cliffs NJ.

Hammer, M. (1990) Don't automate, obliterate, *Harvard Business Review*, Jul–Aug.

Hammer, M. and Champy, J. (1993) *Reengineering the Corporation*, HarperCollins, New York.

Harrington, H.J. (1991) *Business Process Improvement*, McGraw-Hill, New York.

Jain, P., Liu, J. and Wagner, S. (1994) *Transformational approach to business process redesign*, AAAI BPR Workshop, May.

Jordan, N. (1963) Allocation of functions between man and machine in automated systems, *Journal of Applied Psychology*, **47**, 161–5.

Kerr, S. (1975) On the folly of rewarding A while hoping for B, *Academy of Management Journal* (USA), **18**, 769–83.

LeClair, S.R. (1982) IDEF the method, architecture the means to improve manufacturing productivity, *SME Technical Paper MS82–902*, Society of Manufacturing Engineers, Dearborn MI, USA.

Leavitt, H.J. (1962) Unhuman organizations, *Harvard Business Review*, **40**.

Lumsdaine, E. and Lumsdaine, M. (1995) *Creative Problem Solving –Thinking Skills for a Changing World*, McGraw-Hill, New York.

March, J.G. and Simon, H.A. (1958) *Organisations*, Wiley, Chichester.

Maull, R.S., Childe, S.J., Bennett, J., Weaver, A.M. and Smart, P.A. (1995) *Different types of manufacturing processes and IDEF$_0$ models describing standard business processes*, Working Paper WP/GR/J95010–6, University of Plymouth, UK.

Meredith, J.R. (1981) The implementation of computer based systems, *Journal of Operations Management* (American Production and Inventory Control Society), 2(10).

Miller, E.J. and Rice, A.K. (1967) *Systems of Organization – The Control of Task and Sentient Boundaries*, Tavistock, London.

Osborne, A.F. (1963) *Applied Imagination*, Charles Scribner & Sons, New York.

Parnaby, J., Johnson, P. and Herbison, B. (1987) Development of the JIT-MRP factory control system, in *Proceedings of the 2nd International Conference on Computer Aided Production Engineering (CAPE)* (ed. J.A. McGeough), Edinburgh University/Institute of Mechanical Engineers, Edinburgh.

Sandras, W.A. (1989) *Just-in-Time: Making it Happen – Unleashing the Power of Continuous Improvement*, Oliver Wight Ltd Publications, Essex Junction VT, USA.

Simon, H.A. (1960) *The New Science of Management Decision*, Harper & Row, New York.

Skidmore, S., Farmer, R. and Mills, G. (1993) *SSADM Version 4 Models and Methods*, NCC/Blackwell, Oxford.

Skinner, W. (1985) *Manufacturing: The Formidable Competitive Weapon*, Wiley, New York.

von Oech, R. (1983) *A Whack on the Side of the Head*, Creative Think, USA.

Weaver, A.M. (1995) A model based approach to the design and implementation of computer aided production management systems, PhD Thesis, University of Plymouth.

Weaver, A.M., Maull, R.S., Childe, S.J., Smart, P.A. and Bennett, J. (1995) The development and application of a generic 'order fulfilment' process model, *Proceedings of the Third International Conference on Computer Integrated Manufacturing*, Singapore, 8–11 July.

Wortmann, J.C. (1989) Towards an integrated theory for design, production and production management of complex, one of a kind products in the factory of the future, in *ESPRIT 89, Proceedings of the 6th Annual ESPRIT Conference, Commission of the European Communities, Brussels.* Dordrecht, pp. 1089–99.

Case studies

The following case studies are included to provide the reader with the opportunity to think through some of the ideas presented earlier in the book in relation to the needs of particular companies. The three cases presented here are all based on real life, each bringing together real situations from a range of companies. The cases have been written for the purposes of the exercise, and do not reflect the true position in any of the companies from which they were originally drawn.

CS.1 RIGIBORE LTD

CS.1.1 THE COMPANY

Rigibore is a small company based in West Cornwall. Its main products are boring bars which are sold to a wide range of industrial customers throughout the world. The company has sales outlets in the USA and Canada and sells direct to Europe.

CS.1.2 THE PRODUCTS

The company produces a range of tools which are used in metal cutting. The principal product is a wide range of boring bars.

A boring bar is a tool which is used in the machining of an internal bore in a component. The simplest kind of boring bar is similar to a drill, but generally without flutes and with removable cutting tips. Boring bars are made in a variety of different sizes to allow different holes to be machined. The range of standard products allows the customer to select from various standard shanks (ISO tapers etc. to fit various machines), different lengths and diameters, and different types and numbers of holders for tungsten carbide inserts. The holders can be either a pocket into which the insert can be clamped, or a 'Rigibore Unit' which allows fine adjustment of the cutting tip.

Non-standard boring bars are made to customers' special requirements for the machining of holes with internal features, such as shoulders or chamfers. These have to be designed individually, although they are made from standard bar blanks and Rigibore Units. A typical bill of materials is shown in Fig. CS.1.

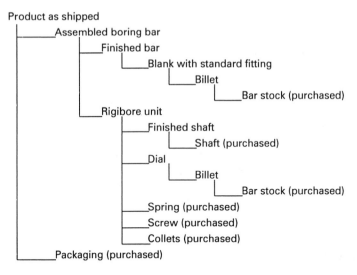

Figure CS.1 Bill of materials.

Rigidity of the boring bars is a key requirement of the customers who order from Rigibore. General purpose boring bars (which Rigibore does not produce) are slender so that they can be used for bores of different sizes. Rigibore offers the ability to use a specific tool for any bore, which can therefore be of a cross-section almost as large as the bore it produces, and as short as possible. These features allow the maximum possible rigidity, and therefore the greatest accuracy of the bore produced.

By using a tool specifically made for the job, it also becomes possible to machine the bore in a single cutting feed, thus eliminating pilot drilling and reaming in many cases. Improved rigidity and accuracy are required in industries such as aerospace, while the elimination of operations and time saving may be the first priority of volume industries such as the automotive industry.

CS.1.3 DESIGN

The design of standard bars is simply a matter of varying the dimensions of the boring bar within the limitations shown in the catalogue. Thus the customer is able to design the product required for a particular machining application.

Special products are designed by the company's engineers according to the customers' special requirements. This involves determining the correct number and angular orientation of the various cutting units and the required clearance for swarf removal. The company has recently created a computer package which will allow sales engineers to design the tool at the customer's premises and then to transmit the data back to the factory in a form suitable for immediate post-processing into programs for the machines in the factory to produce the finished bar.

CS.1.4   SALES

Customer orders are generally for small quantities, typically up to fifty tools. Customers typically buy new tools for one of three reasons:

- *Setting up a new production line.* This generates the largest orders, which may be for some time in the future. However, such orders do not arrive regularly.
- *As a spare to replace a worn or broken tool.* This generates a rush order, since the tool may be required for immediate use.
- *In jobbing manufacture, to produce a feature for which the planning engineer decides a special tool is required.* This may occur when the workpiece is already ordered in the customer's factory, so any lead time at Rigibore extends the lead time for the customer's customer.

CS.1.5   THE FACTORY

The factory comprises a small workshop of approximately fifteen machine tools. The machines are independent of each other and include a range of machining and turning centres, including five- and seven-axis machines, and various others including CNC and manual milling centres, lathes and grinders. There is also an assembly area and various ancillary equipment.

Production of bars is divided into several main stages, and stocks are held at each stage. This means that a customer's order can be satisfied very quickly by finishing stock items and assembling them as required.

Products can be made in a very few operations since many of the machines can combine several features in one operation. The most advanced machine is used for special products and can complete most bars in one operation.

Parts are sent outside for painting and heat treatment, which may take up to three weeks.

Lead times for most parts are between four and five weeks, including shipping.

CS.1.6 THE CONTROL SYSTEM

The company holds stock of finished products in common sizes in its three locations (UK, USA, Canada). Stocks are also held of bar blanks with fittings, Rigibore Units, and all purchased items.

All production is either to finish bars to a customer order or to replenish stock items which have been consumed.

When a customer order is received, it is entered in the sales and accounts computer. The computer works out what stock will be used up and allocates this stock to the order. A second run repeats this to establish the lower level parts to replenish the stock items used. This produces a replenishment list which contains all the items which the company must make or buy. This is then adjusted manually (to provide convenient quantities) and then purchase orders are raised and the manufactured items are produced in the factory. The list is held in the factory office, where the production operators call to find out what job to produce next. The orders are prioritized manually according to the awareness of the production manager of the various customer delivery requirements. The system does not provide any capacity planning, so it is difficult to anticipate overloads. Night shift working and overtime are used as required.

For each customer order, an engineer produces a works instruction sheet which describes the operations required to produce the product from stock items. The works instruction sheet forms the documentation for final inspection. For special products produced in one operation on the seven-axis machine, this requires little more than the production of the program (which is automated). The required stock items are taken and used to satisfy the order.

The general manager wishes to improve the company's competitiveness by reducing the time spent in production of works instruction sheets, and is at the same time considering changing the way in which orders are processed.

QUESTIONS

1. Describe the company's competitive priorities. How can the management of production contribute to achieving these requirements?
2. Produce an outline of the functions which the company's CAPM system must fulfil.
3. What possible solutions would be worthy of investigation for this company?

CS.2   FLY FISHING REELS

Fly Fishing Reels employs 140 people in the south of England. They are the major UK manufacturer of reels for fly fishing. The company was formed five years ago when a previous fishing reel business on the same site went into receivership.

In the hands of the receiver, the company was formed with just six employees from the previous company. Over the five-year period it has grown considerably to its present size, now one of the three major manufacturers of fly reels worldwide. The two main competitor companies are in the USA and Japan.

FFR have a good reputation for quality, and both the 'feel' and the performance of their products are often superior to their competitors'. Some of their models are more expensive than the competition, although discounting and special prices are used from time to time.

Over one hundred different styles of reels are manufactured, although the products fall into three categories. One range is made from plastic, a proportion are manufactured from die-cast aluminium, and the premium range is machined from solid material. In each range there are several variations of reel, housing and attachment foot, multiplied by a wide variety of paint finishes, drilling patterns for weight reduction, and various mechanisms which provide alternative combinations of ratchet and friction control.

The premium products compete on the basis of style, appearance and feel, together with high reliability, low weight and high strength. These sell to professional and 'serious' sportsmen who are prepared to pay a high price for the best equipment. For the majority of the products, the selling price is an important additional competitive factor. Most of the products are sold through retailers and wholesalers, who tend to order products in quantities of at least five or ten for premium products and up to 1000 for other models. Sales – and even retail outlets – may be lost if deliveries are not made on time. Purchasers in shops like to see and feel the product before they buy, and are not usually willing to wait for stock to arrive.

Production of the die-cast products is in several stages, including casting, machining, painting, subassembly and assembly. First, the larger parts (the spool halves, the body and the attachment foot) are die-cast. Next, the die-cast parts must be finish machined. Each casting may be used in a range of products, the variation being introduced by machining, especially drilling. The parts are next painted, as a surface finish, and then pad-printed to provide visual differences between parts which will be used for the different models, to give each model a distinctive appearance. A pre-assembly stage is used to fit together the reel halves, using a rotary riveting process, and to rivet the attachment foot to the frame. At final assembly, the completed product is put together using the main components and whatever minor parts, such as spindle, pawls, springs and ratchet wheels, are needed to produce the required mechanical action, and the finished product is packed for sale. Most of the minor parts are produced in-house.

The premium products follow a similar route, except that die-casting is replaced by machining from solid.

The factory is arranged along process lines, with separate areas for die-casting and fettling, finishing and polishing, machining and NC machining, turning, grinding, painting and assembly.

The assembly area is made up of subassembly, dealing with spool assembly and straightening, and four lines in final assembly. These are rows of benches where the successive stages of the assembly operations take place. Some of the benches are fitted with riveting machines and some with presses. There is no conveyor between the benches, the workpieces being passed along by hand.

A master production schedule is produced with the aid of an MRP-type system. This allows customer orders to be entered, and they are grouped together by type to provide batches of a few weeks' requirements. The MPS shows the quantities obtained by grouping, not the original orders. No formal capacity planning is used to produce the MPS, but the workloads generated by the different product types are very well understood by the production manager, who has considerable experience and very detailed knowledge. He is also able to identify products which use the same components, so that works orders and purchase orders can be released according to the production manager's knowledge of the current state of the factory and the stores. There is thus a considerable degree of flexibility to deal with any problems or delays which may arise. This can mean that the MPS becomes inaccurate, which has the effect of the MRP system suggesting orders which have already been raised manually, or not suggesting orders for items which are to be made earlier than planned. This can happen when the production manager brings forward a batch to keep a particular part of the factory busy. In order to keep production and ordering under control, the production manager has to check all purchase orders before they are released, and can amend quantities as appropriate. He can also use his knowledge of the minute-to-minute situation to release orders manually to the factory and suppliers in response to customer orders the moment they are received, even before these have been registered in the MRP system.

Batch sizes tend to be dictated by the painting operation. The paint equipment requires about three hours to change from one job to the next. The spray head is held on an automatic reciprocator, which requires adjustment for each different item, since the work-holding fixtures are of different sizes. It is also necessary to exchange paint tanks and wash through the paint hoses and nozzles when changing colour. Six different colours are used. The painting equipment is always very heavily loaded, and if smaller batches were allowed, the extra setting time would mean that the current workload could not be processed.

One problem in the MRP system in this company is that minimum order quantities are used for all items. A requirement for a single unit of a high-level item such as a finished body assembly may be increased to a minimum quantity, such as 100, which is felt to be an appropriate batch size for the paint-spraying operation. This then cascades down the bill of materials with the effect of ordering enough components (spool, body, attachment foot, rivets etc.) to manufacture the 100 items, even though there may be enough of some of them in stock to build the single item which is required for the customer. The production manager must therefore pay close attention to the orders being raised by the MRP system.

Trial kitting is used from time to time to determine which order to release next to the factory. This uses the MRP system to show whether the components required to assemble a particular order are in fact in stock. If they are in stock ahead of time, this allows the production manager to release the job to be produced early. If the items are not in stock, action can be taken to expedite them as quickly as possible. Items which are expected to be in stock can sometimes be found to be short, since most of the items are common to several products and some could have been drawn from stores to fulfil an order which was brought forward. This is especially true of items in the unpainted stage. A considerable amount of energy is required of the production manager to keep abreast of the situation and to expedite those orders which can be produced. Nevertheless, it is often impossible to ship some orders on time, and the order value of late orders has in one instance represented an amount equal to an eighth of the annual turnover.

A characteristic of the product which makes this problem worse is the nature of the alloy used for the die-castings. This is a corrosion-resistant aluminium which is difficult to die-cast and also difficult to machine. Minute voids and cracks sometimes appear in the castings, especially around the circumference of the circular items, which may be uncovered when the surface is machined but often not discovered until it is painted. The paint provides a good quality finish for the customer but also reveals any defects in the surface beneath. The number of faulty castings which are discovered in this way has been up to 40% of a batch. This necessitates drawing extra stock and running a new batch through all the machining operations and painting, which involves considerable disruption to production and even more problems for the production manager. Steps taken to deal with this problem (besides the engineering investigations to improve the material or the die-casting process) have so far included adding a scrap allowance to all batches before casting, and holding a safety stock of cast items. The amounts of extra stock needed are difficult to quantify, and there is no way to be sure that all the items held in stock are perfect until they are painted.

At busy times of year, assembly limits production. The larger batches can sometimes take several days to pass through the subassembly stage, and even longer for final assembly. This can mean that the total lead time for the two assembly operations can be around two weeks. Despite the marketing of the product being worldwide, sales are mainly to the Northern hemisphere and there is still a considerable seasonal fluctuation of demand. Orders for the months July to December can be half the volume of the orders received for the other six months. Combined with the long assembly lead time, this means that production is often to a forecast of demand with a build-up of stock being produced in the lighter months. This stock is mostly at the machined stage.

The company is particularly susceptible to the cost of aluminium. Recent increases in the cost of the raw material have led to a price rise.

QUESTIONS

1. Do you feel that the CAPM system in use in this company is providing the production manager with all the information needed to take decisions?
2. Does the CAPM system cause any problems?
3. What factors are limiting the company's sales?
4. What factors are limiting the company's cash flow and profits?
5. If you were to advise the company, what problem areas would you wish to deal with first?
6. How would the company's situation be seen by each of the following experts?
(a) Eliyahu Goldratt
(b) Shigeo Shingo
(c) John Burbidge.

7. How flexible does the company appear to be to deal with the changing mix of customer orders? Does flexibility need to be improved? If so, how?

CS.3 CARPARTS LTD

Carparts have called in your consultancy company to help improve their competitive position. After an initial round of interviews, your colleagues have collected the following information from Carparts staff. You should:

- understand the existing system and identify any problems
- consider the company's objectives for improvement of the system
- design an improved system to meet the company's objectives.

CS.3.1 BACKGROUND – INTERVIEW WITH THE MANAGING DIRECTOR

Carparts employs approximately 160 people manufacturing automotive spares for vehicle manufacturers, automotive spares distributors and retailers. It is owned by a multinational company which owns over 200 subsidiaries.

Six years ago, Carparts moved to a new purpose-built factory which houses all the production, research and administrative resources. It is situated in the south-west of England, whereas the majority of the company's customers in the car industry are based in the Midlands. A very small proportion of the company's sales are for the export market, although it is attempting to increase sales to Europe.

The company produces over 300 different products, which are size and specification variations to cover over 3000 different types of vehicle. The on-site warehouse is capable of holding over 2 million product units, with an average stock level of over 1.4 million units.

The company has a turnover of approximately £15 million per annum, and in the last two years has made losses totalling over £3 million.

The present Managing Director took over nine months ago. The target given to the MD by the parent company was to regain profitability and increase the company's market share by 20% within two years. During the past nine months the work force has been reduced by 20% to its current level of 160 employees, and ISO 9000 accreditation has been awarded.

The MD believes that to regain profitability the company must improve to compete on an equal level to its competitors. A key problem that the MD is very keen to tackle is the time taken from receipt of the order from a customer to the customer receiving the goods. All the company's major UK competitors can deliver goods within 24 hours of the order being placed, while Carparts takes on average 48 hours. Distance is claimed to be a major disadvantage. Many of Carparts' competitors are based in the Midlands and the north of England.

CS.3.2 INVESTIGATIONS

Initial analysis set the boundary for the study and identified the activities involved in the process of order fulfilment. The boundary identified was to include all the activities, people and systems involved with the flow of a customer order from receipt of the customer order in the company to the moment the customer receives the products ordered.

Most of the activities in the order fulfilment process are conducted by staff from four departments, Sales, Production (especially Production Control), Warehousing and the Export Office. Other areas of the company are also involved at various stages.

The managers of the four departments have each been interviewed, with the aim of understanding:

- the activities carried out within the department
- the links between the department and the rest of the company
- the strategy and performance measures of the department as perceived by the relevant manager
- any problems believed to exist within the department.

These issues were not covered comprehensively in the limited time available.

Sales manager

The sales department is the first point of contact for a customer, and is where all orders are received. The department has seven members on site and five full-time representatives covering separate regions of the UK. There are also ten agents for the company in Europe who are paid on a commission-only basis.

The sales manager acts as the sales representative for the local area as well as managing the function. He is regarded as the rising star within the company and is keen to take on new ideas and constantly pushes the sales function to lead new company initiatives. He is prepared to get involved with everything and takes pride in being able to carry out any activity within the function, which means he takes part in all kinds of activities.

Performance of the sales department is measured on value of orders placed during a four-week cycle. The sales department communicates with all other functions: stock information is supplied by the warehouse function via the computer system; customer credit information is supplied by the accounts department on request; promotions on products and discounts to customers must be authorized by the senior management team.

Orders can be received in a variety of formats, including by telephone from either a sales representative or the customer, or by fax, telex or mail on either a standard order form which is given to customers or the customer's own order form. Orders received by telephone are transferred to a standard order form. Stock availability information is not generally given to customers even if they order by phone. All customer orders follow the same procedure.

Once the order has been received, the sales clerk finds the customer's details using an index card system and adds them to the order form. A credit check is carried out, which involves checking, with the on-line accounts system, the level of credit already extended to a customer, external creditworthiness checks in some cases and identifying whether the customer is entitled to receive a discount. Any export order is referred to the export office to ensure that payment guarantees are available and to determine the currency exchange rate.

The credit authorization and the discount are noted on the order form, then the availability of the required items is checked.

If the customer is not sufficiently creditworthy for the full order to be accepted, usually because of credit extended already, the customer is notified and attempts are made to reach a mutually acceptable agreement with the customer, which may involve accepting an order for a reduced amount, or payment before delivery. All changes to the order due to credit problems must be authorized by the sales manager.

The availability of items from stock is established by the sales function staff using on-line access to stock records updated by the warehouse staff.

For each item in the order, if the required quantity of the ordered item is available in stock, this is recorded on the order form and the order quantity for the item is added to a 'picking list' for the order. The level of available stock is adjusted. When all the items in the order have been checked, the picking list is keyed into the computer system. It is then authorized and released, when it is printed out and sent by internal mail to the warehouse. The picking list tells the warehouse staff the quantities of items required, the date required and customer details, so that the items can be picked out.

When the ordered quantity of any item cannot be met in full, the shortage is marked on the order as not available and a 'back order' is created. A separate back order is created for every item type. The back orders are passed to production control via the internal mail. A copy is kept in the sales department until the items are produced and become available in stock. The back order is then processed as a new order, and all the checks are made again.

Once the items on the picking list have been picked by the warehouse, the completed picking list is sent back to the sales department and an invoice is printed. Invoices are printed in a daily print run each afternoon. The invoice is added to the shipment for delivery to the customer.

Export manager

The export department consists of three staff: the manager and two clerks. There are two major areas to their work. The first area is to ensure that the export of orders goes smoothly, which involves checking the import regulations of the country, ensuring that the correct methods of packaging and transportation are used for the order, and producing documentation in a language acceptable to the customer. The second area is to act as interpreters between other members of the company and foreign customers when there are language problems.

The manager carries out an active role in every activity in the department. The performance objectives for the manager are to ensure that there is a minimum delay in export orders reaching their customers and to ensure that the warehouse uses the correct packaging. This is done by sending a packaging note to the warehouse.

Production control manager

The production control department consists of three employees who are responsible for controlling production and for purchasing raw materials.

Production is driven by a month-by-month demand forecast, which is generated by the production control manager in close consultation with the sales manager and the MD. The forecast is combined with a daily shortage report produced from the back orders generated by the sales department and the current stock levels of products. The document which covers all three is described within the company as the 'master plan'. The production control department uses the master plan to calculate the quantities of different items to be manufactured. Item quantities generated by the forecast always have a requirement date at the end of the month, item quantities that are required to replenish stocks have a required date of the current date plus product lead time, and item quantities that are required for back orders have the current date. Using the master plan, production control staff determine the items required and adjust the quantities to predetermined economic batch quantities.

The master plan is reviewed daily by the production control manager. For each individual item type that is shown for the forthcoming week, a single 'work book' is produced by the senior production controller. One work book can represent a manufacturing order for many thousands of units of one type. In generating the work books, raw material stocks are allocated to the production of those items as far as they are available.

Work books for which there are insufficient raw materials are passed to the other production controller to order more raw materials. This involves the issue of a purchase order and establishing an agreed delivery date with the supplier. The work book is not released to the shop floor until a goods inwards note is received to indicate that the material has arrived and the raw material can be allocated.

The shop floor supervisors have full control of the work books and allocation of resources to get the items specified on the work books by the date required. Production is by a series of flow lines, which can produce an order of a thousand items in a few hours, according to the type of item. The production lines can be changed over from one product to another in a time ranging from less than an hour to a few hours, according to the nature of the changeover.

Warehouse manager

The warehouse manager has a staff of eight who carry out all activities from the storage of finished products to the picking and packaging of customer orders. The manager takes sole responsibility for the booking of transport services for the delivery of orders, but the company does no actual deliveries and owns no vehicles. There is an attempt to keep the collection of orders for delivery to once a day, as extra collections incur extra charges from the carrier. There are often requirements to deliver urgent parts by special courier, but the bulk of the orders leave on the single daily collection. Some large customers send their own transport to collect the goods.

The warehouse staff act on receipt of the picking list, which they stamp with the date received. They assess the picking list in terms of priority and special requirements, such as for special packaging or courier delivery, and place the lists in order of priority.

Members of staff collect the picking list of highest priority, identify the locations of the items on the picking list and collect the items to be packaged. If items on the picking list are not available, the quantity of items to be delivered is amended. This happens if the items are not actually available despite information on the computer system to the contrary. Once all items from the picking list are collected together ready for packaging, the completed picking list is returned to the sales office, where the invoice will be produced.

When the invoice is received, the warehouse staff match up the documentation with the goods, which will by then have been packed. The order is now ready for despatch to the customer at the next collection time or for collection by special courier. When the order leaves, its despatch is registered on the computer system and the accounts department is notified.

QUESTIONS

1. How do Carparts believe they win customer orders? Does the company have appropriate systems to satisfy this requirement? Does the company pay attention to any other competitive requirements?
2. Evaluate the company's improvement strategy.
3. Use a diagram to describe the order fulfilment process at Carparts. Describe any problems which hinder the company's competitiveness.
4. Consider the role of each department in the order fulfilment process. Are there any problems at the interfaces between departments?
5. What issues would you address to improve the company's competitiveness in the short term and in the long term? What should the priorities be?
6. How would you redesign the process in the short term to deal with the issues identified?

Appendix A
Flow charts

Many different types of flow chart are used, since they are so common that many companies have made up their own rules.

In the simplest form of flow chart, activities are shown as boxes and the sequence in which they are performed is indicated by arrows which lead from one box to the next. The boxes are often arranged in a column so that the arrows lead vertically down the page.

A more sophisticated flow chart is based upon a standard form described by the American National Standards Institute. This provides a set of different shaped boxes to add more meaning to the chart. The boxes which indicate activities can be of various shapes to indicate useful operations (rectangular box), movement of items (arrow-shaped box), inspection (circle), or delay (D-shaped box). Symbols are also available to show input information (slanted box), information which appears on a screen (shaped like a cathode ray tube) or on paper (rectangle with a torn-off bottom edge). Cylinders are often used to indicate computer files.

The flow through the chart can take various paths if required, by the use of decision boxes which allow branching, so that the subsequent activity could be one of various alternatives according to the decision taken by the reader for the matter in mind. The reader can see how other activities would be used in the case in which other decisions would be taken. Decision boxes are diamond-shaped (rhombus), with the entry point being at the top and various options being indicated by arrows from the other corners. Where the decision is a simple 'yes' or 'no', the 'yes' is generally downwards and the 'no' to one side.

Other symbols may be used to indicate the beginning and end of the flow, to break the flow between pages, to add notes, etc. A simple example is the operation of MRP as shown in Chapter 4 (Fig. 4.1).

In a further variant of the flow chart, the page is laid out in vertical columns, each of which indicates individuals or departments who carry out the activities. The flow chart is arranged so that each activity lies in the column indicating by whom it is performed.

Flow charts are very useful for describing in detail a sequence of events relating to an object which can be thought of as flowing. If the sequence of events is long, the flow chart can become very large. A flow chart used by a manufacturing company recently to indicate the flow of customer orders was over thirty feet wide and four feet high, comprising over 250 boxes. It is difficult to divide up such charts into smaller sections to be dealt with separately because the chart does not show what information is available at any point, so the activities and the need for them can only be understood by looking at the chart as a whole. It is very easy for the wood to be hidden by the trees.

The boundary of the system being considered is not clear, since inputs and output may appear at any point in the diagram. The symbols for input and output information are often used in different ways or not used at all, so it is hard to be sure that a diagram is complete. It is difficult to show the circumstances surrounding an activity, the arrow simply showing which activity or item is next. Influences such as orders, schedules and regulations which affect activities are not easy to show.

Another difficulty with flow charts is that the lack of complexity means that anyone can draw one. This has the effect that people tend to draw them differently, so they may be difficult for other people to understand. Various software packages are available which include an application for drawing flow charts.

Appendix B
ICAM Definition
Method (IDEF$_0$)

IDEF$_0$ is the activity modelling technique produced by Sof'Tech for the USAF's Integrated Computer Aided Manufacturing project of the 1970s (LeClair, 1982). It was developed with the aim of understanding or defining the environment within which a particular system operates, and it is the first of a series of related techniques. The name IDEF$_0$ is a shortened form of 'ICAM *Definition* Method' and the technique itself is an adaptation of SADT (Marca and McGowan, 1988). IDEF$_0$ has been accepted as a standard by the USA's National Institute of Standards and Technology (NIST, 1993).

IDEF$_0$ uses a very simple graphical technique with only one shape of box, but with a more rigorous set of rules to ensure the correct interpretation of the diagrams. It is very closely related to the concepts of system theory described earlier.

The basic IDEF$_0$ box is a rectangle which represents an activity. The output created by the activity is shown by an arrow which leaves the box from the right-hand side. The input(s) which are converted into the output(s) enter from the left-hand side of the box, thus providing a notional flow from left to right. The activity box thus represents the boundary of a system, in which an activity or a process transforms inputs into outputs.

The conditions necessary for the system to operate, and which affect the way the system operates by triggering or prohibiting its activity are shown by arrow(s) entering the box from the top, known as 'controls' or 'constraints'. As the control or constraint enters the box from outside, it can also be described as an input, but it is shown at the top to indicate that it is the special kind of input which regulates the activity.

The means by which each activity is performed, such as by a person, a department, a computer program or a machine is shown by an arrow which points up to the bottom of the box. This is called the 'means' or the 'mechanism'.

The four types of arrow are shown in Fig. B.1. The control can be thought of as starting, controlling or stopping an activity performed by the mechanism.

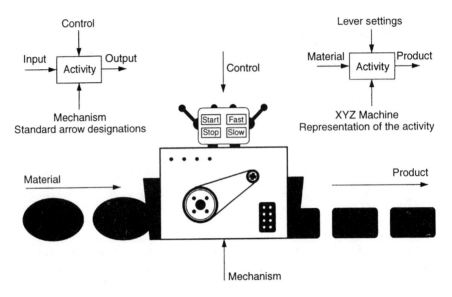

Figure B.1   IDEF$_0$ arrow designation and representation of an activity.

IDEF$_0$ activity boxes are arranged on the page from top left to bottom right as shown in Fig. B.2. This arrangement allows the arrows to be followed easily. Arrows above the diagonal are either controls (vertical) or outputs (horizontal), and those below are inputs (horizontal) and mechanisms (vertical). This allows minimum crossing of arrows where outputs from one box become controls to another.

A minimum of three and a maximum of six boxes are recommended on each page. This makes each page simple enough to be understood. If a system were investigated and a larger number of activities was found, these would be dealt with by grouping them into between three and six sub-systems, each of which could then be represented by another diagram of three to six boxes. The grouping (or 'decomposition', if it is done in reverse) provides a structure which helps the reader to understand the system which the model describes. For example, if 23 activities were identified, it would create a very complicated diagram if they were all shown on one sheet of paper. Instead, they could be grouped as five boxes on a summary sheet which would be called A0. The boxes would be numbered A1 to A5. The detail contained in the first box, A1, would be shown on a separate sheet of paper called A1, which might consist of three activities. These would be numbered A11, A12 and A13.

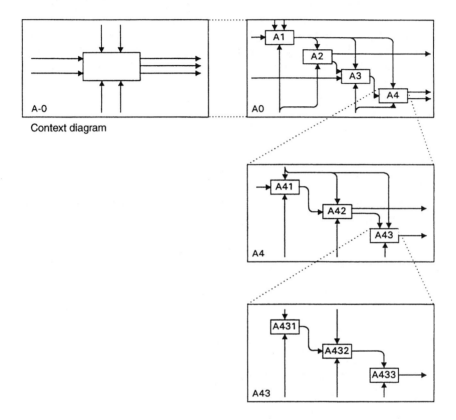

Figure B.2 Schematic showing grouping and decomposition in $IDEF_0$.

Thus,

diagram A0 shows A1, A2, A3, A4 and A5
diagram A1 shows A11, A12 and A13
diagram A2 shows A21, A22, A23, A24 and A25

etc.

Every box shows the contents of a box on a higher level diagram, so the environment within which each activity operates can be identified, as shown in Fig. B.2. If the reader wishes to understand a control on box A432, it will either come from one of the other boxes on the page or from the top edge of the page. If it comes from one of the other boxes, say A431, then there may be a more detailed diagram of A431 which will show how it is produced. If it comes in from the top of the page, this indicates that it enters box A43 from somewhere which will be revealed by looking at the diagram of A4. Thus every item affecting any box can be traced either to an activity which generates it, or eventually to a top-level diagram which shows the arrow entering from the environment outside the entire system. The entire system is shown in relation to the outside world on a 'context diagram', which is exactly the same as showing box A0 with no detail, as it would appear on a higher level diagram, if there was one. The context diagram is given the special number A-0, by convention.

Six important rules of $IDEF_0$ are the following:

- Three to six boxes per page, except for the single-box context diagram.
- Each activity must be described by an imperative verb. An imperative verb is a verb in the form used for commands, such as 'process orders', 'deliver products', 'buy material'. The use of an imperative verb ensures that the activity name describes what happens. The *I-test* can be used to check if a term is an imperative verb, thus 'I process orders', 'I test subassemblies', but not 'I accounts office'.
- Every activity must have at least one control (otherwise it would never happen!). Several controls may act on one activity: for instance, a policy statement or guideline on how the activity must be performed and a piece of information whose arrival starts the activity.
- All arrows must be labelled. Since all arrows represent things, it is vital to know which things.
- Arrows may join or split. Where an arrow shows information or a mechanism which is used in more than one process, the arrow may split. Where different items go to the same activity, it may be convenient to join the arrows. This is a way of removing clutter from the diagram and must not be used instead of an actual activity, such as assembly, which must be shown with its controls.
- Arrows may be combined at higher levels and decomposed at lower levels. Since all arrows must be traceable on higher level diagrams, the number of arrows on the top level diagrams can cause confusion. Combining arrows can allow detailed information at lower levels to be represented by a general term at a higher level.

Thus the arrows 'customer name', 'customer address', 'customer account number' and 'customer invoice number' may need to be distinct at a low level, but at a high level where they all take the same path they may be represented by a single arrow such as 'customer details'. Great care must be taken to ensure that the generalized arrow on the high-level diagram can be seen to represent the same items as the detailed arrows on the low-level diagram. This can be achieved by careful use of names, and by the use of supplementary notes if necessary.

The strength of the $IDEF_0$ technique is that it can show the level of detail required for any analysis, while still relating each activity or sub-system to the context in which it must fit. Project teams working with $IDEF_0$ can operate at whatever level of detail is required.

A design project can focus on the contents of a particular box (whether at a high or low level) and redesign it completely, as long as the new sub-system will still fit into the box. This is in line with the systems concept of working with a system as a whole while taking account of its environment and its relationship to wider systems.

$IDEF_0$ shows how parts of systems interact so that the effects of changes in one area do not cause unexpected problems in other areas. Most importantly, it allows a group working on a problem to explore their understanding and to form a consensus view, so that they can work together to develop the new system.

REFERENCES

LeClair, S.R. (1982) IDEF the method, architecture the means to improve manufacturing productivity, *SME Technical Paper MS82-902*, Society of Manufacturing Engineers, Dearborn MI, USA.

Marca, D.A. and McGowan, C.L. (1988) *IDEF$_0$/SADT Business Process and Enterprise Modeling*, Eclectic Solutions Corp., San Diego CA, USA.

NIST (1993) Integration definition for function modeling (IDEF$_0$), *Federal Information Processing Standards Publication 183*, National Institute of Standards and Technology, USA.

A41 Consolidate order

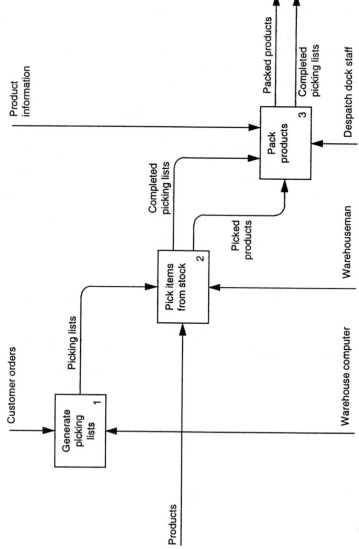

Figure B.3 Example IDEF$_0$ diagram.

Appendix C
Paired comparisons

Paired comparisons is a technique which structures a decision process concerning many possible solutions and many criteria for judgement.

Criteria are those attributes of a solution which require judgements to be made and may include attributes such as cost, training required, speed of operation and user-friendliness. The criteria can be compared and ranked against each other to determine their relative importance, using a paired comparison matrix as shown in Box C.1. Each criterion has a row and a column, and each possible pair is compared in a square in the matrix. In each square a '1' is entered if the row criterion is thought more important than the column criterion. In the example, 'speed' is thought more important than 'cost'. The rows are totalled and the criteria are ranked. They are then given a weighting which is felt to be appropriate.

Box C.1   Paired comparisons for criteria weighting.

	Criterion						
Criterion	a	b	c	d	Total	Rank	Weight
a Cost	–	0	0	0	0	4th	1
b Speed	1	–	0	0	1	3rd	2
c Easy to understand	1	1	–	1	3	1st	5
d Simple to use	0	1	1	–	2	2nd	3

The possible alternatives can now be compared with each other in the light of each criterion, using a separate matrix for each criterion, each with a row and column for each solution alternative. A matrix for the criterion 'speed' is shown in Box C.2. Each solution alternative has a row and a column, and each possible pair is compared in a square in the matrix. In each square a '1' is entered if the row solution is thought to win out over the column solution in respect of the particular criterion in question. Thus a modification to the existing software (solution b) is expected to offer better speed than using a logbook (solution a). The rows are totalled and the totals are multiplied by the weight of the criterion.

Box C.2   Paired comparison matrix for solution alternatives in respect of one criterion (speed of use, weight 2).

Solution	Solution					Total	Weighted
	a	b	c	d	e		
a Logbook	–	0	0	0	0	0	0
b Modify existing software	1	–	0	1	1	3	6
c New package	1	1	–	1	1	4	8
d Card file	1	0	0	–	0	1	2
e Adapt spreadsheet	1	0	0	1	–	2	4

Once all the matrices have been completed, so that the solutions have all been compared against each other in the light of each criterion, the weighted scores can be totalled up. The solution with the highest score wins.

Once a result has been obtained from the paired comparison technique, it is customary to alter the weightings until the desired solution is proved to be the best. This is a legitimate group activity and may be essential to achieving consensus.

Index

Page numbers appearing in **bold** refer to figures and page numbers appearing in *italic* refer to tables.